TECHNICAL REPORT

Analysis of Government Accountability Office Bid Protests in Air Force Source Selections over the Past Two Decades

Thomas Light, Frank Camm, Mary E. Chenoweth,
Peter Anthony Lewis, Rena Rudavsky

Prepared for the United States Air Force

PROJECT AIR FORCE

The research described in this report was sponsored by the United States Air Force under Contract FA7014-06-C-0001. Further information may be obtained from the Strategic Planning Division, Directorate of Plans, Hq USAF.

Library of Congress Cataloging-in-Publication Data

Analysis of Government Accountability Office bid protests in Air Force source selections over the past two decades / Thomas Light ... [et al.

   p. cm.

 Includes bibliographical references.

 ISBN 978-0-8330-5091-5 (pbk. : alk. paper)

1. Defense contracts—United States—Evaluation. 2. United States. Air Force—Procurement—Evaluation. 3. Letting of contracts—United States. I. Light, Thomas, Ph. D. II. Title.

KF855.A87 2012

358.4'16212—dc23

2011022737

The RAND Corporation is a nonprofit institution that helps improve policy and decisionmaking through research and analysis. RAND's publications do not necessarily reflect the opinions of its research clients and sponsors.

RAND® is a registered trademark.

Published 2012 by the RAND Corporation

1776 Main Street, P.O. Box 2138, Santa Monica, CA 90407-2138

1200 South Hayes Street, Arlington, VA 22202-5050

4570 Fifth Avenue, Suite 600, Pittsburgh, PA 15213-2665

RAND URL: http://www.rand.org/

To order RAND documents or to obtain additional information, contact Distribution Services: Telephone: (310) 451-7002;

Fax: (310) 451-6915; Email: order@rand.org

Preface

This technical report presents the findings of analyses of data on U.S. Air Force acquisition protests submitted to the Government Accountability Office (GAO) over the preceding two decades. It is part of a larger study of the Air Force's recent experience with bid protests that performed case study analyses of recent successful, high-profile bid protests and identified various lessons learned. RAND Project AIR FORCE (PAF) undertook this effort at the request of Gen Donald Hoffman, as former military Deputy to the Assistant Secretary of the Air Force for Acquisition (SAF/AQ); Lt Gen Mark D. Shackelford, SAF/AQ; and Roger S. Correll, then–Deputy Assistant Secretary of the Air Force for Contracting (SAF/AQC). These sponsors asked PAF to identify specific changes that the Air Force can make to both its source selection policies and processes for complex acquisitions so as to minimize incidents that often lead to successful protests. This research was conducted as part of a project titled "Air Force Source Selections: Lessons Learned and Best Practices," conducted during fiscal year (FY) 2009 within PAF's Resource Management program.

To analyze the Air Force's bid protest experience, we used federal administrative databases of Air Force procurement and protest activity. We summarized trends and developed statistical models to identify factors contributing to protest activity and outcomes. The findings of these analyses should be of interest to those with direct responsibility for source selections and bid protests, as well as to policymakers working in these domains.

The companion documents for this report are:

- *Government Accountability Office Bid Protests in Air Force Source Selections: Evidence and Options,* Frank Camm, Mary E. Chenoweth, John C. Graser, Thomas Light, Mark A. Lorell, Rena Rudavsky, and Peter Anthony Lewis (DB-603-AF).
- *Government Accountability Office Bid Protests in Air Force Source Selections: Evidence and Options—Executive Summary,* Frank Camm, Mary E. Chenoweth, John C. Graser, Thomas Light, Mark A. Lorell, and Susan K. Woodward (MG-1077-AF).

RAND Project AIR FORCE

RAND Project AIR FORCE (PAF), a division of the RAND Corporation, is the U.S. Air Force's federally funded research and development center for studies and analyses. PAF provides the Air Force with independent analyses of policy alternatives affecting the development, employment, combat readiness, and support of current and future air, space, and cyber forces.

Research is conducted in four programs: Force Modernization and Employment; Manpower, Personnel, and Training; Resource Management; and Strategy and Doctrine.

Additional information about PAF is available on our website:
http://www.rand.org/paf.html

Contents

Figures

Tables

Summary

When an offeror in an Air Force source selection believes that the Air Force has made an error that is large enough to change the outcome of the source selection, the offeror can file a protest with the Office of General Counsel of GAO. Following its review, GAO can suggest a course of remediation to the Air Force, if it agrees that a significant error has occurred and that the error has the potential to change the source selection outcome. GAO cannot compel the Air Force to follow its suggestion, but if the Air Force fails to do so, GAO must report this noncompliance to Congress. However, the Air Force almost always follows GAO suggestions when it sustains a bid protest.

The Air Force experienced an average of 93 protests a year on contract awards between FY 2000 and FY 2008. GAO sustained on average three of these protests each fiscal year, recommending that the Air Force make significant changes in how it had conducted the source selections associated with these contract awards. However, simultaneously, the Air Force addressed 30 protests per year by preemptively engaging in corrective action. Corrective actions can involve such steps as reevaluating proposals, reopening evaluations and giving offerors an opportunity to adjust their proposals, changing the offerors included in the source selection, rewriting the request for proposals and starting the source selection from scratch, or even canceling the acquisition as a whole. If a protester accepts such action, there is no further need for the GAO to review the protest. If the protester rejects the corrective action, the protest continues through the GAO review process.

During the summer of 2008, the Military Deputy, SAF/AQ and SAF/AQC, asked PAF to analyze the Air Force's performance in GAO bid protests. Using federal administrative databases, PAF conducted detailed analyses of Air Force protest activity and outcomes over the past two decades. The aim was to determine which factors are correlated with increased protest activity across Air Force contracting units as well as to identify characteristics associated with particular protest outcomes. This technical report describes these analyses and supports another PAF report, Camm et al. (2011a), *Government Accountability Office Bid Protests in Air Force Source Selections: Evidence and Options*, which places the findings reported here in a broader policy setting.

Research Approach

To inform our sponsors about Air Force experience with GAO bid protests, RAND researchers undertook the following analytic tasks:

1. reviewed documents on general Air Force source selection policies and processes, comparable policies and processes elsewhere in the Department of Defense, and others elsewhere in the federal government
2. reviewed other studies of bid protest activity, including analyses conducted by the Air Force itself, the Congressional Research Service, and GAO
3. assembled data on Air Force protests contained in the Protest and Congressional Tracking System (PACTS) database; as part of an effort to understand these data better, we spoke with the personnel responsible for updating and maintaining PACTS
4. gathered and assembled for analysis information on Air Force procurements contained in the Individual Contracting Action Report (form DD350) and the Federal Procurement Data System–Next Generation (FPDS-NG)
5. performed tabulations to characterize the Air Force's experience with bid protests both over time and across different contracting units
6. estimated statistical models that rely on the PACTS, DD350, and FPDS-NG databases and that control for a variety of factors to draw inferences about the determinants of Air Force protest activity and outcomes.

Findings

Although broad criticism of the Air Force has focused on a few recent protests sustained by GAO, these incidents represent only a small portion of the protests that the Air Force has experienced. Between FY 2000 and FY 2008, the Air Force awarded over 133,000 contracts and experienced 836 protests. It offered corrective action in 273 cases, or 33 percent of all protests. Meanwhile, it ultimately suffered sustained protests in only 29 cases, or just 3 percent of all protests.

Presumably, the Air Force offers a corrective action when it believes that GAO will sustain a protest and suggest a corrective action at least as onerous as the one the Air Force itself could proffer. Conversely, a protester accepts a corrective action early when it believes that GAO will offer nothing significantly better. Therefore, an early corrective action is likely to prevail if both the Air Force and the protestor believe they will derive greater value from avoiding a full-fledged intervention.

To put protests in perspective, it is useful to express them in terms of the total number of contract awards the Air Force makes. The number of protests as a percentage of total contract awards has fallen fairly steadily from about 1.7 percent in FY 1995 to 0.5 percent in FY 2008. Sustained protests are so unusual that they hardly register relative to the total number or value of contract awards. In fact, so few sustained protests have occurred that it is impossible to discern any meaningful trend in them. On the other hand, the Air Force has offered corrective actions in noticeable numbers. Throughout the 1990s, it offered corrective actions in about 0.3 percent of contract awards. Yet, from FY 2001 to FY 2008, the percentage fell fairly steadily, ending well under 0.2 percent of contract awards. On the whole, all of these trends point to improvement over time.

Conclusions derived from simple tabulations of the protest and procurement data provide important insights into the Air Force's performance with bid protests. However, formal statistical analysis that accounts simultaneously for multiple factors affecting protest activity can provide greater clarity into protest activity. Our statistical analysis led us to find that the likelihood of bidders pursuing protests with GAO has been declining over time at a rate of between 8 and 9 percent per year, after controlling for other factors. This further supports the finding that the Air Force's experience with bid protests has been improving over time.

Looking across the Air Force, we found that contracting centers having technical activities, with about 5 percent of the total spending, had substantially better experience than the Air Force as a whole, with 45 percent fewer protests than normal for the Air Force after controlling for other factors. This finding is consistent with simple tabulations of protest and procurement activity across Air Force contracting centers. Contracting centers having technical activities may have qualitatively different types of contracts, with lower baseline protest risk. Alternatively, they may have contracting cultures more attuned to conducting complex source selection evaluations.

Our statistical analysis also suggested that the number of protests tends to increase less than proportionately when the number of contract awards rises (holding all other factors, including spending, constant).

A second line of statistical analysis, which sought to identify factors correlated with different protest outcomes, yielded additional insights. Namely, our statistical analysis of the determinants of protest outcomes supported the general observation that, over the 1990s, there was a steady increase in the probability that a protest would lead to the Air Force offering a corrective action, while simultaneously controlling for other trends that were occurring at that time. We did not find a significant correlation between protest outcomes and the commodity or service being acquired by the Air Force. We found some evidence that different types of contracting centers as well as the basis for protest are associated with different probabilities of engaging in corrective action or facing a sustained protest; however, it is difficult to statistically differentiate between any two types of contracting centers or causes of protest.

Acknowledgments

This project would not have been feasible without the cooperation, assistance, and goodwill of many people knowledgeable about Air Force source selection and the GAO bid protest process. We give special thanks to Roger S. Correll, Randall Culpepper, and Maj Brett Kayes of SAF/ACQ, who helped us frame the analysis and provided valuable documents and introductions to relevant contacts. Even a cursory examination of this technical report will quickly reveal how much we benefited from the foundational empirical analysis by Maj Kayes (Kayes, 2008). We also thank Sarah Dadson, Olalani Kamakau, Michael J. Maglio, Sharon Mule, and Pamela Schwenke for their valuable insights.

Others in the office of the Secretary of the Air Force gave us important insights into recent Air Force experience with bid protests and the ongoing development of the Acquisition Improvement Plan. They include Lt Col Carole Beverly; Lt Col Jon C. Beverly, Office of the Executive Action Group of the Assistant Secretary of the Air Force for Acquisition (SAF/AQE); Kathy L. Boockholdt, Office of the Air Force Acquisition Center of Excellence (SAF/ACE); Blaise J. Durante, Deputy Assistant Secretary for Acquisition Integration, Office of the Assistant Secretary of the Air Force for Acquisition (SAF/AQX); Maj Jon Dibert (SAF/AQXD); Patrick M. Hogan, director of SAF/AQXD; James A. Hughes, Jr., Deputy General Counsel for Acquisition, Office of the Air Force General Counsel (SAF/GCO); Robert Pollock (SAF/ACPO); and Col Neil S. Whiteman, chief of the Air Force Commercial Litigation Directorate (AFLOA/JAQ).

RAND colleagues Elliot Axelband and Paul Heaton provided excellent comments and feedback on the technical report. Mark Lorell and Jack Graser, who participated in the broader project with which this research is associated, provided helpful discussion and feedback. Laura Baldwin oversaw this work and took an active interest in it throughout. Susan Gates helped us gain access to and understand the personnel data that she had used in the past. Judith Mele similarly helped us interpret the contracting data that she had worked with in the past. Megan McKeever was always ready to provide helpful administrative support.

Abbreviations

AFB	Air Force base
AFLOA/JAQ	Air Force Legal Operations Agency/Commercial Litigation
AFMC	Air Force Materiel Command
ALC	air logistics center
CSAR-X	Combat Search and Rescue Recovery Vehicle
DD350	Individual Contracting Action Report
DoD	Department of Defense
FPDS-NG	Federal Procurement Data System–Next Generation
FY	fiscal year
GAO	Government Accountability Office
GDP	gross domestic product
HQ	headquarters
IDIQ	indefinite delivery/indefinite quantity
KC-X	Aerial Refueling Tanker Aircraft
MAJCOM	major command
NA	not available
OB	other base
OLS	ordinary least squares
OSD	Office of the Secretary of Defense
PACTS	Protest and Congressional Tracking System
PAF	Project AIR FORCE
PB	projection base
PC	product center

PIID	procurement instrument identifier
R&D	research and development
RFP	request for proposal
SAF/ACE	Assistant Secretary of the Air Force, Acquisition Center of Excellence
SAF/AQ	Assistant Secretary of the Air Force for Acquisition
SAF/AQC	Deputy Assistant Secretary of the Air Force for Contracting
SAF/AQE	Office of the Executive Action Group of the Assistant Secretary of the Air Force for Acquisition
SAF/AQX	Deputy Assistant Secretary for Acquisition Integration, Office of the Assistant Secretary of the Air Force for Acquisition
SIAD	Statistical Information Analysis Division
TCC	technical contracting center
USAF	United States Air Force

CHAPTER ONE

Introduction

Bid protests made headlines when the Government Accountability Office (GAO) sustained protests in the Air Force's Combat Search and Rescue Recovery Vehicle (CSAR-X) program and Aerial Refueling Tanker Aircraft (KC-X) program source selections. The protests in the CSAR-X program in fiscal year (FY) 2007 so disrupted Air Force planning that resulting delays helped lead to the cancellation of the program. Meanwhile, the protests in the KC-X program in FY 2008 delayed the high-priority, phased tanker recapitalization effort by three years, with the source selection remaining uncompleted until the beginning of 2011. As a consequence, the Office of the Secretary of Defense (OSD) chose to remove the Air Force's authority to oversee this source selection for a period of time. In response, the Air Force leadership has made a concerted effort to understand why these protest sustainments occurred and how to avoid similar sustainments in the future.

The Protest and Congressional Tracking System (PACTS) database, which tracks Air Force experience with GAO bid protests, records protests that have occurred over the period FY 1991–2008. Between FY 2000 and FY 2008, the Air Force experienced 836 protests, or about 93 a year. This database is a natural place to look for insights into protest activity and outcomes. This technical report documents analysis of PACTS data, as well as two other databases that were used to place the contents of PACTS into a broader setting—the Individual Contracting Action Report (form DD350) and the Federal Procurement Data System—Next Generation (FPDS-NG). It supports another Project AIR FORCE (PAF) report, which places the findings reported here in a broader policy setting.[1]

Summary of Findings

Although public and congressional criticism of the Air Force has emphasized recent protests sustained by GAO, they are only part of a much larger picture. To put protests in perspective, it is useful to express them in terms of the total number of contract awards the Air Force makes. The number of protests as a percentage of total contract awards has fallen fairly steadily from about 1.7 percent in FY 1995 to 0.5 percent in FY 2008. Sustained protests are so unusual that they hardly register relative to the total number or the value of contract awards. On average, between FY 2000 and FY 2008, GAO sustained one protest for every $20 billion the Air Force spent on procurements. So few sustained protests occurred that it is impossible to discern any

---

[1] Frank Camm, Mary E. Chenoweth, John C. Graser, Thomas Light, Mark A. Lorell, Rena Rudavsky, and Peter Anthony Lewis, *Government Accountability Office Bid Protests in Air Force Source Selections: Evidence and Options*, Santa Monica, Calif.: RAND Corporation, DB-603-AF, 2012a.

trend in them. On the other hand, the Air Force offered corrective actions in noticeable numbers. Throughout the 1990s, it offered corrective actions in about 0.3 percent of contract awards. From FY 2001 to FY 2008, the percentage fell fairly steadily, ending well under 0.2 percent of contract awards. All of these trends point to steady improvement over time, resulting in a need to adjust fewer than 0.2 percent of the source selections associated with contract awards by FY 2008.

In this study, we report more detailed tabulations of data from administrative databases that support these findings. We also perform statistical analysis to understand the key factors correlated with protest activity and outcomes, controlling for other factors that may have attributed to protests. The statistical analysis suggests the following additional findings:

- After controlling for other factors, the likelihood of receiving a protest has been declining at a rate of 8 to 9 percent per year between FY 1994 and FY 2008. When attention is restricted to the likelihood of receiving a protest that undergoes a merit review by GAO or receives a corrective action, the rate of decline falls slightly but exceeds 6 percent per year in most of the empirical specifications we tested. This finding is consistent with general trends observed in the protest record but also controls for other factors that are likely to be important, such as the amount of contracting activity being undertaken by the Air Force.
- Protest activity tends to increase as the total number of contract awards rises, but the increase is less than proportional. That is, if the number of contract awards doubles, the analysis suggests that we would expect to see less than a doubling in protest activity, holding all other factors equal including expenditure. These results are robust across functional forms and specifications.[2]
- When assessing protest performance across Air Force contracting centers, we found that centers that conduct test and evaluation activities[3] tend to perform better than other Air Force contracting centers, holding constant such other important factors as spending and the number and nature of contracts. The acquisition environment in these centers (e.g., personnel, policies) may provide useful lessons for the rest of the Air Force.[4]
- In terms of predicting protest outcomes based on the characteristics of the underlying protest (i.e., whether a protest is sustained by GAO or results in a corrective action by the Air Force) rather than on the number of protests received, our statistical findings tend to be less robust, but some findings are worth briefly noting. Namely, assessment of protest outcome determinants supports the casual observation that over the 1990s there was a steady increase in the probability that a protest received a corrective action, while controlling for other trends that were also occurring at that time. This said, we did not find a significant correlation between the commodity or service being acquired by the Air Force and whether the protest resulted in an unfavorable outcome for the Air Force. There is some evidence that, when taken as a whole, different contracting centers are associated with different probabilities of experiencing sustained rulings by GAO or Air Force correc-

---

[2] Statistical models with three alternative covariate structures are estimated using ordinary least squares (OLS) and Poisson and negative-binomial count regressions.

[3] In our analysis, we call these centers "technical contracting centers."

[4] That does not mean other parts of the Air Force can simply adopt policies or personnel types observed in technical contracting centers. Typically lessons observed in one organizational setting must be carefully adapted before application in another to reflect differences in mission, priorities, organizational context, resources, and so on.

tive actions; however, it is difficult to statistically differentiate between any two centers. In some specifications, the basis for the protest proved to be statistically correlated with negative protest outcomes.

Although such analysis is useful for analyzing protest activity and outcomes, data limitations do not allow us to explore the explanations of these findings more deeply. As a result, we have conducted case studies of the two most recent and significant source selections that resulted in successful protests against the Air Force, namely, CSAR-X and KC-X. Interested readers are encouraged to review Camm et al. (2011a) for this analysis and further exploration of the root causes of and potential remedies for negative Air Force source selection protest outcomes.

Road Map for This Technical Report

This technical report documents the findings that we developed from our analysis of protest and contract data. Chapter Two describes trends in GAO protests, both over time and across types of Air Force contracting centers. Chapter Three provides findings of statistical analyses of broad patterns in the Air Force's historical experience with bid protests. Chapter Four concludes with a summary of the findings and a discussion of additional analysis that may be useful. The appendix describes the data sources used and associated cleaning and manipulation tasks.

The discussion here presumes that the reader has a simple understanding of how the Air Force's source selection process and GAO's bid protest process work. Chapter Two in Camm et al. (2011a) provides a primer on these processes that may be useful for understanding the findings reported here.

Bid Protest Patterns and Trends

This chapter reports quantitative evidence on the Air Force's experience with bid protests over the last two decades. Data from the Air Force's PACTS database allow us to document basic patterns in the characteristics of protests that arose and the outcomes that followed. Data from DD350 and FPDS-NG allow us to put such protest information in a broader context by identifying any aggregate factors beyond the specific source selection processes that might have affected the Air Force's track record. The data used in this analysis, as well as our efforts to clean and process the data for analyses, are described in greater detail in the appendix. The tabulations provided in this chapter set the stage for Chapter Three, where statistical modeling techniques are employed to identify key factors that are correlated with protest trends and outcomes while simultaneously controlling for a variety of variables.

When an offeror files a protest, GAO assigns designated protests one or more B-numbers (e.g., B-123456.2, B-123456.3). A record is created in PACTS for each protest B-number that the Air Force tracks. GAO may assign multiple B-numbers to an acquisition protest if multiple offerors file protests or if the protested actions encompass distinct issues as defined by GAO. B-numbers associated with the same source selection will generally have the same root B-number (i.e., B-123456 in B-123456.2) under the GAO system, although a few exceptions to this do exist.

Past analyses of protest trends have typically relied on counts of B-numbers, instead of on unique root B-numbers.[1] Because a protest can be associated with more than one B-number, these analyses typically overstate the true number of protests. Counts of unique root B-numbers are more indicative of the actual number of acquisitions that have been protested. In this technical report, therefore, tabulations of the number of protests correspond to counts of unique root B-numbers unless specifically noted otherwise.

Basis for Recent Protests

As described in Chapter Two of Camm et al. (2011a), errors in the source selection process can occur at a variety of different steps and in a variety of different ways. Table 2.1 presents

---

[1]   See, for example, Congressional Research Service, *GAO Bid Protests: Trends, Analysis, and Options for Congress*, February 11, 2009a. This is discussed in Congressional Research Service, *Report to Congress on Bid Protests Involving Defense Procurements*, April 9, 2009b.

Table 2.1
Frequency of Reasons for Protest (FY 2000 to FY 2008)

Protest Basis	Number of Protests	Percentage of Protests
Faulty evaluation	522	62
Faulty RFP	146	17
Faulty sourcing decision	150	18
Faulty treatment of offerors	71	8
Other	29	3
NA	2	0
Total	836	

NOTES: Frequency counts are based on root B-numbers and calculated from PACTS. Some protests with multiple B-numbers are associated with multiple protest reasons, causing the sum of protests by protest basis to equal more than the total number of protests. NA indicates that the protest basis was not reported in PACTS.

a summary view of the frequency of different reasons for protests raised between FY 2000 and FY 2008.[2] The protest categories presented in Table 2.1 have been defined as follows:

- **Faulty evaluation:** The evaluation criteria employed by the Air Force are considered inconsistent with Air Force rules, regulations, policies, procedures, or the performance parameters or trade space, as defined in the request for proposal (RFP).
- **Faulty RFP:** The RFP was flawed in a way that the protester believes unfairly disadvantaged its ability to respond. This includes the adoption of restrictive specifications or requirements that appear to favor one or more parties' existing capabilities.
- **Faulty sourcing decision:** The Air Force's proposal restrictions or selection of a winner was flawed. This includes issues stemming from small business rules, sole source requirements, cancellation of a solicitation, etc.
- **Faulty treatment of offerors:** The Air Force treated the protester's offer unfairly or in a way that disadvantaged it and is counter to Air Force rules, regulations, policies, or procedures. This includes faulty determinations that a bid is out of the competitive range, leading an offeror's proposal to be excluded from consideration, or having improper, closed discussions with certain bidders.
- **Other:** A catchall category for other protest reasons.

Table 2.1 shows that over the period studied, protesters asserted that errors occurred throughout the source selection process. None occurred in the requirements determination phase, because this is beyond GAO's acquisition-oriented jurisdiction. But a significant number

---

[2] As with the data found in many administrative databases, those coming from fields in PACTS that are not used frequently to inform decisions tend to require considerable processing before being useful for analyses. "Reason for protest" appears to be one such field. Patterns of reporting appear to shift over time as the personnel recording the data change, suggesting some ambiguity in the labels someone working today might have applied if he or she had recorded the data 15 years ago. As a result, in some of the tables and figures presented here, we focus on the post-FY 1999 period because the data were not available or reported inconsistently during the pre-FY 2000 period. That said, we believe that these data present a rough idea of how reasons for protests occur across steps of the source selection process.

of protests were associated with issues arising before the Air Force even issued an RFP. These tended to involve asserted errors in how the Air Force organized a competition or classified a protester in a competition.

Sixty-two percent of protesters claimed that errors occurred during the proposal evaluation phase, e.g., as the result of a failure to evaluate a proposal the way the RFP indicated, use of unreasonable basis to justify an evaluation, or failure to document the Air Force's position in real time during evaluation. Analysis of recent sustained protests indicates that these problems have been the dominant errors specifically highlighted in formal GAO sustained decisions.[3]

Protest Outcomes

For the purposes of our review, we differentiate between protests in terms of their GAO outcome. Each B-number that receives a merit review is either sustained or denied. A protest is considered sustained overall if any B-number associated with the protest is sustained by GAO. For instance, if a protested acquisition (i.e., a unique root B-number) receives eight distinct B-numbers and only one B-number is sustained, then the entire protest is classified as sustained for tabulation purposes. However, if a protest has at least one B-number that receives a merit review, yet no B-number receives a sustainment, we classify the protest as being denied. If a protest has at least one B-number that is either sustained or denied, we characterize it as a merit protest for the purposes of our tabulations.

Protests that do not receive a merit review are either dismissed by GAO or withdrawn by the protesting party. If one or more B-numbers associated with a protest are dismissed, we classify the protest as dismissed for the purposes of our analysis.[4] Protests that are not sustained, denied, or dismissed by GAO are classified as withdrawn.

In many instances, a protester will withdraw its complaint after the Air Force takes voluntary corrective action. However, in the event the protestor does not withdraw, GAO will dismiss the protest if the corrective action renders the pending protest academic, i.e., moot. We classify a protest as having a corrective action protest if one or more B-numbers associated with the protest received a preemptive corrective action. We emphasize, for the point of clarity, that it is almost never the case in a corrective action protest that the voluntary corrective action itself is being protested, since this is a mutually agreed upon remedy between the Air Force and aggrieved party.

Figure 2.1 summarizes the outcomes of protests made over Air Force acquisitions between FY 2000 and FY 2008. During this time, the Air Force experienced protests in 836 acquisitions.[5] Meanwhile, GAO conducted a merit review in only 201 of these acquisitions—or about a quarter of the total. The remainder were either dismissed by GAO or withdrawn by the protester. Preemptive corrective actions are associated with 43 percent of the dismissed protests and 30 percent of the withdrawn protests.

---

[3] For evidence from FY 2006–2008, see Brett N. Kayes (Capt, USAF), "Air Force GAO Protest Trend Analysis," briefing, Washington, D.C.: SAF/AQC, updated September 19, 2008, Charts 9–12.

[4] Eleven protests in PACTS reported as closed are considered dismissed for the purposes of these tabulations.

[5] As noted above, protested acquisitions can be associated with multiple B-numbers. The 836 protested acquisitions noted here correspond to 1,113 B-numbers.

Figure 2.1
Air Force Protest Outcomes (FY 2000 to FY 2008)

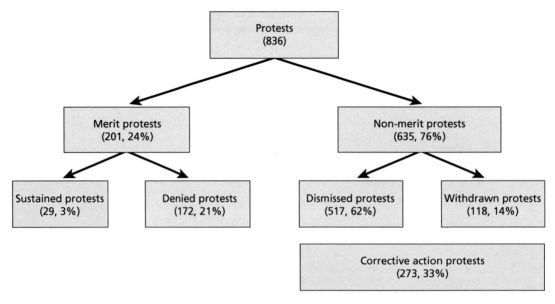

RAND *TR883-2.1*

Two points are worth noting. First, 64 percent or 534 out of 836 protests overall did not ultimately necessitate a substantive response from the Air Force, while consuming valuable administrative resources until mooted. However, 273 protests, or about a third of the total, did lead the Air Force to voluntarily offer preemptive corrective action that a protester accepted. Second, of the 201 protests for which GAO conducted a merit review, it sustained just 29 cases, or 3 percent of protests overall.

All protests impose some administrative burden on the Air Force, and merit reviews in particular cause time delays in the acquisition process that can complicate planning, but these costs pale in comparison to those imposed by voluntary corrective actions or GAO-mandated efforts. Consequently, remediating the root causes that give rise to these latter two scenarios is more likely to yield benefits for the Air Force than seeking to abate frivolous claims.

Causes of Protest Outcomes

Table 2.2 provides a breakdown of the types of corrective actions taken by the Air Force.[6] Given the prevalence of perceived evaluation errors in the protests filed with GAO, the similar dominance of corrective actions that reevaluate proposals is not surprising. This type of corrective action is likely to impose less cost on the Air Force than the others detailed in Table 2.2. For instance, reopening discussions allows offerors to adjust their proposals, potentially in ways that can significantly complicate the Air Force's evaluation of the changes. Repeating a solici-

---

[6]   Note that Table 2.2 tabulates corrective actions by B-number rather than by root B-number. "Unknown" applies to cases in which PACTS says that a corrective action occurred but does not specify the form of corrective action. We are cautious about our confidence in the accuracy of PACTS data on corrective actions. However, viewed at this level of aggregation, we believe that they offer useful insights into relative numbers of different kinds of corrective actions.

Table 2.2
Corrective Actions Taken (FY 2000 to FY 2008)

Corrective Action	Number	Percentage
Cancel solicitation	50	13
Repeat solicitation	95	25
Reopen discussions	24	6
Reevaluate	146	39
Other	58	15
Unknown	5	1
Total	378	100

NOTES: Tabulations are based on B-numbers. Percentages do not sum to 100 because of rounding.

tation adds still more administrative cost and delay. Finally, outright cancellation can have a crippling effect on planning, particularly when a solicitation is associated with time-sensitive recapitalization efforts.

GAO dismissed over 60 percent of Air Force source selection protests during the FY 2000–2008 period. As Table 2.3 clearly demonstrates, the basis for those dismissals points to the persistent presence of protesters who did not understand the GAO protest process when they filed a protest. In about half the protests that occurred, either GAO lacked jurisdiction over the issue in question, the protester lacked standing before GAO, the protester could not state a legal basis for the protest that met GAO's standards, or the protester used GAO's protest process incorrectly. Otherwise, GAO dismissed the vast remainder as academic—cases no longer of interest as a result of preemptive resolution between the Air Force and the protesting party.

It is clear that such unsound protests impose administrative and legal costs on the Air Force, but these burdens are likely to be mainly an irritant when compared to the expenditures and schedule slippage associated with sustained protests and corrective actions. Therefore, the Air Force will undoubtedly deliver the greatest value to both the taxpayer and the warfighter

Table 2.3
Reasons Protests Were Dismissed by GAO (FY 2000 to FY 2008)

GAO Dismissal Basis	Number	Percentage
No longer relevant (academic)	299	45
Lack of GAO jurisdiction	59	9
Lack of standing	35	5
Lack of legal basis	66	10
Failure to use GAO process properly	160	24
Other	47	7
Total	666	100

NOTE: Tabulations are based on B-numbers.

by addressing core factors that lead to substantive, costly, and unforeseen changes in its acquisition endeavors, rather than by focusing its attention on after-the-fact administrative matters.

For the remainder of this technical report, we differentiate between protests in terms of their GAO outcome, focusing on (1) those that involved a merit review or that prompted the Air Force to take a corrective action (what we call substantive protests), (2) those merit protests that are sustained, and (3) those protests that prompted the Air Force to take voluntary remediation (corrective action protests).

A More Detailed Look at Trends in Air Force Bid Protests over Time

In this section, we take a closer look at Air Force protest trends over time. Table 2.4 categorizes Air Force protests by type (substantive, merit, and sustained and those that led the Air Force to take preemptive corrective action without a GAO sustained ruling).[7]

Over the FY 1991 to FY 2008 period, GAO reviewed 25 percent of all protests against the Air Force. During FY 2007 and FY 2008, this rate declined to only 16 and 8 percent of protests, respectively, suggesting a positive trend. Meanwhile, approximately 3 percent of all protests (or 12 percent of all merit protests) were sustained each year. Although the sustained protest rate fluctuated somewhat, the number of sustained protests was so low that it is inappropriate to place much weight on these variations.[8]

The trend for corrective actions is less positive. Over the FY 1991 to FY 2001 period, the percentage of protests that prompted preemptive corrective action by the Air Force increased steadily, reaching a high of 39 percent of all protests in 2001. Since then, the corrective action rate has fluctuated between 29 and 36 percent. This rise in corrective actions no doubt accounts for some, but not all, of the decline in the number of merit cases.

To better understand how Air Force procurements have been evolving over time, we further compiled data from the DD350 and FPDS-NG databases on the number of contract awards, payments, and actions, as well as the aggregate number of contracts outstanding over time. These figures for FY 1994 to FY 2008 are summarized in Table 2.5.[9]

Tables 2.4 and 2.5 can be used to put the Air Force's broad experience with bid protests in perspective, focusing on the numbers of sustained protests and corrective actions relative to the number of procurements. In particular, note that during the FY 2000 to FY 2008 period, the Air Force made 14,813 awards per year on average. Over that same period, there were 93 protests per year on average, corresponding to approximately 0.62 percent of all acquisitions. Of these 93 protests, on average only 33 acquisitions led to voluntary corrective action or a sustained ruling by GAO. Taken together, these numbers imply that the Air Force had to

---

[7]   Note that the number of substantive protests tends to be slightly less than the number of merit protests plus the number of sustained protests in Table 2.4. This is because some protests with multiple B-numbers may receive corrective action on some B-numbers and a merit review on others B-numbers causing the protest to be counted as both a corrective action protest and a merit protest.

[8]   The protest counts shown in Figure 2.1 cover only the FY 2000 to FY 2008 period, whereas Table 2.4 provides information for the earlier period, going back to FY 1991. As suggested by the data, the number of protests and corrective actions were changing over the FY 1991 to FY 2008 period, so we focus on the more recent FY 2000 to FY 2008 period for the purpose of characterizing the Air Force's recent protest experience in Figure 2.1.

[9]   We did not attempt to analyze procurement data before FY 1994 because of the data limitations described in the appendix.

Table 2.4
Substantive, Merit, Sustained, and Corrective Action Protests (FY 1991 to FY 2008)

Fiscal Year	No. of Protests	No. of Substantive Protests	Substantive Protest Rate, %	No. of Merit Protests	Merit Rate, %	No. of Sustained Protests	Sustained Rate, %	No. of Protests with a Corrective Action	Corrective Action Rate, %
	a	b	c = b/a	d	e = d/a	f	g = f/a	h	i = h/a
1991	328	116	35	115	35	17	5	2	1
1992	295	90	31	74	25	7	2	17	6
1993	354	115	32	79	22	8	2	40	11
1994	242	95	39	62	26	6	2	43	18
1995	228	89	39	60	26	6	3	31	14
1996	176	66	38	37	21	1	1	34	19
1997	157	62	39	26	17	2	1	39	25
1998	127	62	49	34	27	5	4	36	28
1999	121	58	48	30	25	4	3	28	23
2000	90	49	54	28	31	3	3	27	30
2001	98	54	55	24	24	2	2	38	39
2002	84	45	54	19	23	2	2	28	33
2003	110	63	57	37	34	3	3	34	31
2004	78	43	55	21	27	0	0	28	36
2005	97	57	59	27	28	9	9	35	36
2006	88	43	49	22	25	5	6	27	31
2007	93	41	44	15	16	4	4	28	30
2008	98	35	36	8	8	1	1	28	29
Total	2,864	1,183	41	718	25	85	3	543	19

NOTE: Protest tabulations were developed by RAND using PACTS data and calculated using root B-numbers.

address a problem in only 0.23 percent of source selections between FY 2000 and FY 2008. Consequently, it is apparent that despite considerable challenges with specific source selections, the acquisition system as a whole is hardly "broken" when it comes to meeting regulatory standards of fairness. Moreover, to the extent (albeit imperfectly) that such rules encourage competition, enhance efficiency, or fulfill other publicly defined goals, such broad-based adherence is valuable in its own right. Alternatively said, it is just as important to recognize value generated by properly conducted source selections as costs associated with adjusting an outcome after the fact.

Based on the tabulations presented in Table 2.4 and Table 2.5, Figure 2.2 illustrates the number of protests per 1,000 contract awards for the FY 1994 to FY 2008 period. Notice that the number of protests of any type as a share of total awards fell dramatically.

Table 2.5
Air Force Contracting Activities over Time (FY 1994 to FY 2008)

Fiscal Year	No. of Contract Awards	No. of Active Contracts	No. of Contract Actions	Contract Expenditures (Nominal, $ Millions)	Contract Expenditures (Real, $ Millions 2008)
1994	18,921	19,970	55,202	39,462	53,597
1995	13,212	21,324	59,580	37,077	49,330
1996	13,625	22,286	61,023	39,081	51,025
1997	12,746	21,138	56,157	34,943	44,831
1998	12,000	20,161	55,149	33,654	42,694
1999	12,469	20,408	55,886	35,246	44,066
2000	11,263	19,290	52,491	37,894	46,373
2001	10,789	18,808	50,737	40,282	48,206
2002	12,651	21,079	58,015	47,351	55,763
2003	14,691	23,839	67,701	55,340	63,798
2004	12,716	22,346	87,142	55,140	61,813
2005	13,219	23,234	116,682	55,068	59,739
2006	13,500	23,721	115,538	62,131	65,275
2007	24,357	31,043	106,043	67,688	69,134
2008	20,129	29,701	107,233	63,188	63,188

NOTES: Tabulations are for contracts with at least one transaction amount greater than $25,000. Tabulations are derived from the DD350 database for FY 1994 to FY 2006; tabulations for FY 2007 and FY 2008 are derived from the FPDS-NG. Contract expenditures were converted to FY 2008 dollars using the gross domestic product (GDP) price deflator (Bureau of Economic Analysis, undated).

What caused this decline? Remember, we define a protest that results in a voluntary corrective action or a merit review as a substantive protest. Next, notice that the trend for number of substantive protests as a share of total contract awards fell fairly steadily from FY 1994 to FY 2008 but not nearly as fast as the total number of protests. Therefore, the decline in substantive protests alone can be only part of the story behind the much larger drop. So what, then, accounts for the remainder of the downward movement? Consider further that the difference between total and substantive protests is the number of protests in which the protester likely made a procedural error of some kind. In turn, note that the number of such erroneous protests as a share of total awards fell dramatically until stabilizing after FY 2000, hence accounting for the majority of aggregate movement. From FY 2001 onward, a shift is apparent—improvements in the total number of protests come mainly from reductions in the number of substantive protests.

Still, what has driven down the number of substantive protests? The number of corrective actions per 1,000 contract awards moved around somewhat arbitrarily until FY 2001, before beginning a long downward trend that helps explain part of the pattern of improvement. The remaining component of substantive protests, the GAO merit reviews, necessarily accounts

Figure 2.2
Bid Protests, by Protest Outcome per 1,000 Contract Awards (FY 1994 to FY 2008)

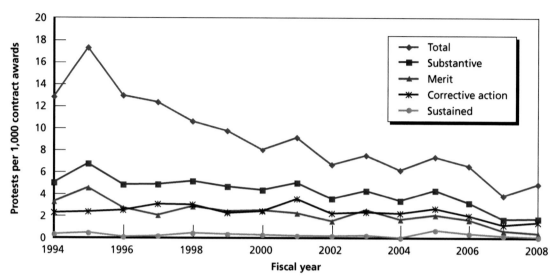

NOTES: This chart is constructed from data in PACTS, DD350, and FPDS-NG. All of these are administrative databases. We have versions of DD350 and FPDS-NG that RAND has significantly cleaned and maintained over time. Viewed at this level of aggregation, we believe the qualitative patterns traced using these data are valid. Alternatively, we could have used the number of competitive contract awards to normalize these outcomes. However, as is explained in the text, because the numbers of total contract awards and competitive contract awards have moved together so closely for the Air Force, the choice denominator has no effect on the qualitative character or policy implications of the patterns that we report here.
RAND TR883-2.2

for the remainder. Indeed, we observe a steady drop in the number of merit protests per 1,000 contract awards.

In contrast, the number of sustainments per 1,000 contract awards is so small that it displays no discernible trend over this time period. In effect, this ratio is constant. However, a constant number of sustained protests in combination with a falling number of merit protests implies that the share of merit protests the Air Force loses has increased over time. We caution that too much focus on this trend obscures three more important points, all of which are positive for the Air Force:

1. The total number of sustained protests per 1,000 contract awards has not risen.
2. As noted above, the number of corrective actions has heavily dominated the number of sustained protests, and the number of corrective actions per 1,000 contract awards has steadily fallen.
3. The number of merit protests per 1,000 contract awards the Air Force must respond to has been falling, reducing administrative costs and schedule slippage.

Figure 2.3 provides insight into the general environment in which the Air Force has been awarding contracts.[10] The number of awards fell through the 1990s as the Air Force drew

---

[10] As an aside, note that because the trends for total and competitive contract awards move together so closely, using the number of competitive awards to normalize protest activity would not yield results that differ qualitatively from those based on the total number of awards.

Figure 2.3
Air Force Contracting Activity over Time (FY 1994 to FY 2008)

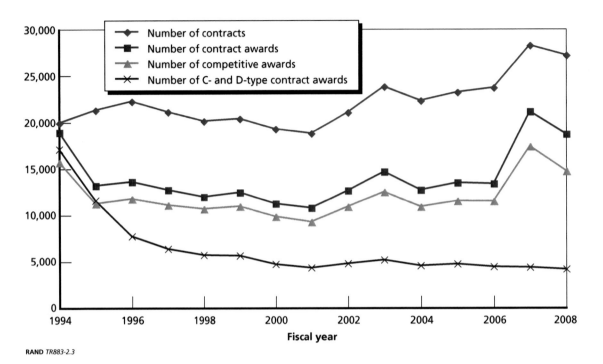

RAND *TR883-2.3*

down its force structure and procurement of new systems to deliver the post–Cold War peace dividend. However, the number of awards then increased as the Air Force both began to rely increasingly on external providers of services (in contrast to its tradition of organic support) and rose to the operational challenges posed by the global war on terror.[11]

The number of C- and D-type contract awards,[12] which tend to be larger and more complex than other contract award types, fell dramatically through the 1990s and then stabilized following FY 2000. Yet, the total number of contracts changed much less, being relatively stable through the 1990s before beginning a steady rise in the new millennium. Consequently, across the full range of years, the Air Force on average handled a falling proportion of high-complexity awards. Of course, it would be desirable if more detailed statistical analyses of the Air Force's experiences with bid protests were employed to control for these patterns of change. The econometric models discussed in Chapter Three attempt to do just this.

---

[11] The Department of Defense completed its implementation of FPDS-NG in FY 2007, effectively replacing DD350 as its primary system for reporting spending. This could potentially change the quality and comparability of reported data from the pre-FY 2007 and post-FY 2006 periods and suggests some caution when interpreting trends for the most recent two years.

[12] C-type contracts are those that are definitive, in that they apply to a fixed quantity of products or services. D-type contracts correspond to those that are unique indefinite delivery/indefinite quantity (IDIQ) awards. These contracts do not specify the exact amount of products or services to be delivered or the time of performance. Delivery orders (for products) and task orders (for services) are issued by the contracting officer to obtain performance under these contracts.

Figure 2.4 displays the level of spending on procurement, shown in constant FY 2008 dollars, as well as the number of contract actions per fiscal year.[13] We observe that aggregate expenditures and contract actions were relatively stable throughout the 1990s, despite the underlying changes then afoot. However, both contract actions and expenditures then rose rapidly through the early 2000s, before stabilizing at a higher level.

If we posit that expenditures or actions are positively correlated with workload, we would expect increasing raw demand on the workforce over the last decade. That said, improved acquisition information systems, as well as other acquisition reforms, might have offset the potential effects of these trends on workload. Simultaneously, changes in the level and composition of the acquisition workforce would also affect the Air Force's response to these trends. Unfortunately, without information on the large, external contractor workforce that supported the organic Air Force acquisition workforce during this time, it is difficult to accurately understand and control for such manpower effects.[14]

Figure 2.4
Air Force Contract Awards and Expenditures (FY 1994 to FY 2008)

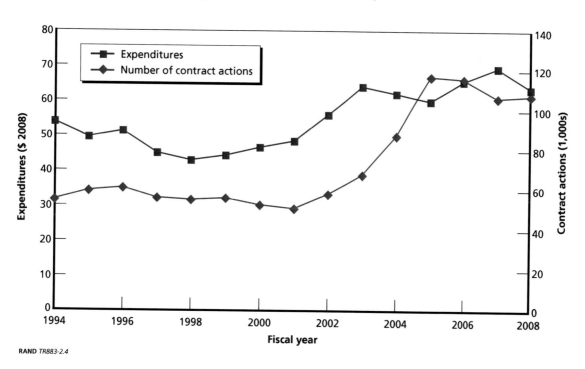

RAND TR883-2.4

---

[13] A "transaction" is the unit of observation in DD350 and FPDS-NG. An action effectively signals a significant change in the level of Air Force funds obligated to a particular contract vehicle. A single contract, therefore, may have many actions both within and across years.

[14] For information on trends in the organic workforce over some of this period, see Kayes, 2008, Charts 29–30.

Protest Activity Across Different Types of Air Force Contracting Centers

Do Air Force experiences with bid protests differ across different parts of the organization? Intuitively, we would expect them to. Suborganization differences likely arise as the Air Force buys different goods and services, uses a multiplicity of contract vehicles and source selection methods, and potentially maintains different contracting cultures across its bases and commands. Further, the level of training and experience of the personnel involved in source selections also differs across contracting centers. Therefore, to examine such hypotheses empirically, we grouped contracting bodies in the Air Force into the following six categories:

- The *headquarters* (HQ) for each Air Force major command (MAJCOM) and Unified Command. The contracting squadrons of headquarters units acquire certain kinds of goods and services that are used by all bases under their organizational umbrella. These contracts tend to be larger and employ more formal source selection activities than those at individual base locations. Examples of headquarters units included here are Langley Air Force Base (AFB), Air Combat Command; Peterson AFB, Air Force Space Command; Scott AFB, Air Mobility Command; Ramstein Air Base, U.S. Air Forces in Europe; Hickam AFB, U.S. Air Forces, Pacific Command; Randolph AFB, Air Education and Training Command; Headquarters Air Force Materiel Command (AFMC); and Cheyenne Mountain, U.S. Strategic Command.
- *Product centers* (PCs) include the Aeronautical Systems Center, Wright Patterson AFB; the Electronic Systems Center, Hanscom AFB; Air Armament Center, Eglin AFB; and Space and Missile Center, Los Angeles AFB. These bodies tend to buy major systems and the services associated with them, using relatively highly trained and experienced workforces.
- *Air logistics centers* (ALCs) include Oklahoma City ALC, Tinker AFB; Ogden ALC, Hill AFB; Warner Robins ALC, Robins AFB; and, before 2001, Sacramento ALC, McClellan AFB; San Antonio ALC, Kelly AFB; and the Aerospace Guidance and Metrology Center, Newark AFB.
- *Technical contracting centers* (TCCs) exist at units and locations that conduct test and evaluation activities, such as Edwards, Eglin, or Kirtland AFBs, or are research and development labs, such as the Air Force Research Lab, which includes Philips Lab, Kirtland AFB; Wright Lab, Wright Patterson AFB; Rome Lab, Griffiss AFB; and Armstrong Lab, Brooks City Base. TCCs employ highly trained and experienced personnel for both test and evaluation and related research and development.
- *Projection bases* (PBs) are defined as individual bases or locations that operate weapon systems having a wartime mission or that can be deployed. Except for the Air Force Materiel Command, bases for each MAJCOM are included. Contracts written by projection bases provide base installation or base operating support. They often have smaller dollar values and require fewer formal source selections. Projection bases operate tactical and strategic aircraft, i.e., fighters and bombers, airlifters, and space assets. Guard and reserve units are included.
- *The other base* (OB) category includes those organizations, units, and locations that either do not have a wartime mission, even though they might provide essential support in the event of an engagement, or typically do not deploy assets. These centers buy a wide variety of goods and services, include complex contractor logistics support services, maintenance

services for individual reparable items, logistics engineering services, supply of consumables and spare parts, and so on. The skills of the workforce in these centers reflect the nature of these purchases. The units comprising this category include Air Education and Training Command, the Human Systems Wing, and all direct reporting units to the Chief of Staff of the Air Force, such as the U.S. Air Force Academy and Andrews AFB.

Table 2.6 provides a summary of protest and procurement activity for each Air Force contracting grouping described above. Notably, product centers account for about half of all Air Force procurement spending over this period. Product centers further have markedly higher sustained protest rates per contract award than the rest of the Air Force, with the exception of the other base category. However, because sustained protest rates remain small relative to voluntary corrective actions, this distinction does not lead to large differences between product centers and the rest of the Air Force when protest activity is measured in other ways. Therefore, it is not surprising that Air Force–wide acquisition outcomes are similar to those in the product centers.[15] Indeed, only outcomes for technical activities and other base activities stand out. Experiences in technical activities are relatively strong in that they experience fewer total protests and negative protest outcomes per contract award; in contrast, outcomes in other base activities are relatively weak. Yet, the reasons underlying this apparent variation are obscure. As a consequence, in Chapter Three, we statistically investigate the key determinants of protest trends among the various contracting centers to clarify these divergent outcomes.

---

[15] The product centers tend to engage in larger dollar-value procurements. As a result, the number of protests per dollar of expenditure tends to be lower for product centers than for other contracting centers.

Table 2.6
Summary of Average Annual Contracting and Protest Activity, by Type of Contracting Office (FY 2000 to FY 2008)

	Air Force Total	ALC	HQ	OB	PB	PC	TCC
Average Annual Contracting Activity							
Contract awards	14,813	3,672	1,792	1,795	4,389	1,458	1,708
Active contracts	23,673	5,473	2,598	3,118	6,531	2,848	3,106
Contract actions	84,620	16,800	11,306	9,597	24,341	14,728	7,847
Contract expenditures ($ billions 2008)	59.3	12.5	4.4	4.0	4.4	30.7	3.2
Average Annual Protest Counts							
Protests	92.9	15.4	11.4	20.9	33.2	8.2	3.7
Substantive protests	47.8	8.7	6.6	10.3	16.3	4.4	1.4
Merit protests	22.3	4.2	3.7	4.6	6.8	2.4	0.7
Sustained protests	3.2	0.9	0.2	1.0	0.3	0.8	0
Corrective action protests	30.3	5.3	3.9	6.3	11.2	2.8	0.8
Protests per 1,000 Contract Awards							
Protests	6.3	4.2	6.4	11.6	7.6	5.6	2.1
Substantive protests	3.2	2.4	3.7	5.8	3.7	3.0	0.8
Merit protests	1.5	1.1	2.0	2.5	1.5	1.7	0.4
Sustained protests	0.2	0.2	0.1	0.6	0.1	0.5	0
Corrective action protests	2.0	1.5	2.2	3.5	2.6	1.9	0.5
Protests per $1 Billion of Expenditures ($ 2008)							
Protests	1.6	1.2	2.6	5.2	7.6	0.3	1.1
Substantive protests	0.8	0.7	1.5	2.6	3.7	0.1	0.4
Merit protests	0.4	0.3	0.8	1.1	1.5	0.1	0.2
Sustained protests	0.1	0.1	0.1	0.2	0.1	0.0	0
Corrective action protests	0.5	0.4	0.9	1.6	2.6	0.1	0.2

NOTES: Protest counts are based on root B-number counts. Contract expenditures are reported in FY 2008 dollars using the GDP price deflator.

CHAPTER THREE

Statistical Analysis of Bid Protest Trends

A variety of techniques can be used to statistically test for factors that have affected the number of protests the Air Force has experienced. These methods use historical data on Air Force experiences to test whether a factor of interest affects protest activity, holding constant other factors that we can control for.

In this chapter, we present the results of two such statistical modeling efforts. In the first, regression models are estimated to understand the factors associated with changes in annual protest counts across types of Air Force contracting centers. The second analysis estimates the probability that a protest is sustained or has a corrective action taken on it as a function of underlying protest characteristics.[1] Taken as a whole, this chapter's results support the positive picture of Air Force experiences with GAO bid protests as outlined in Chapter Two.

Modeling Protest Counts, by Type of Contracting Center

In this section, we model the number of protests by type of contracting center and year as a function of underlying contracting center characteristics. As a first cut, we model this relationship using OLS, where the regression equation takes the form:

$$\ln(y_{it}) = \beta x_{it} + \varepsilon_{it}.$$

In this specification, y_{it} denotes a count of the number of protests that occur in contracting center type i in year t, x_{it} is a set of contracting center covariates, β is a coefficient vector to be estimated via OLS, and ε_{it} is an independent and identically distributed random error term.

The fact that protest counts occur in discrete increments suggests that other modeling specifications may be more appropriate. Indeed, the Poisson and negative binomial regression models have been widely used when working with count data and are well suited for our application.[2] We place greater weight on the estimates derived from these models.

The Poisson distribution implies that the probability of observing y_{it} protests is calculated as follows:

---

[1] Some have argued that factors outside the Air Force acquisition system, like the general state of the economy, might affect the trends examined here. Our analysis considers only factors that can be directly linked to Air Force acquisitions.

[2] For an overview of the Poisson and negative binomial models, see Chapter 25 in William Greene, *Econometric Analysis*, 6th ed., Pearson–Prentice Hall, 2008.

$$\text{Prob}(y_{it}) = \frac{e^{-\lambda_{it}} \lambda_{it}^{y_{it}}}{y_{it}!} \quad \text{for } y_{it} = 0,1,2,3,...$$

where $\lambda_{it} = e^{\theta x_{it}}$, θ is a coefficient vector to be estimated via maximum likelihood, and x_{it} is a set of contracting center covariates at time t as in the OLS model described above.

Given the Poisson formulation, it can be shown that the expected value and variance of y_{it} conditional on x_{it} is equal and related to λ_{it} as follows:

$$E[y_{it} \mid x_{it}] = Var[y_{it} \mid x_{it}] = \lambda_{it} = e^{\theta x_{it}}.$$

Because $E[y_{it} \mid x_{it}] = e^{\theta x_{it}}$, one can interpret continuous covariates that enter the model unadjusted as semi-elasticities and those that enter the model in logs as elasticities.

The assumption of the Poisson model that the mean and variance of y_{it} are equal often does not hold in practice and is criticized as being too restrictive. The negative binomial model is a generalization of the Poisson model, which allows the variance of the distribution of y_{it} to vary with a dispersion parameter k, which is estimated along with a vector δ of coefficients via maximum likelihood. Under the negative binomial specification, the variance of y_{it} is related to the expected value of y_{it} via the following formulas:

$$E[y_{it} \mid x_{it}] - \mu_{it} = e^{\delta x_{it}}$$
$$Var[y_{it} \mid x_{it}] = \mu_{it} + k\mu_{it} = e^{\delta x_{it}} + ke^{\delta x_{it}}.$$

The likelihood function for the negative binomial model is given by

$$\text{Prob}(y_{it}) = \frac{\Gamma(y_{it} + 1/k)}{y_{it}! \Gamma(1/k)} \left(\frac{k\mu_{it}}{1 + k\mu_{it}} \right)^{y_{it}} \left(\frac{1}{1 + k\mu_{it}} \right)^{1/k} \quad \text{for } y_{it} = 0,1,2,3,...$$

where $\Gamma(\cdot)$ is the gamma function.

One can test the validity of the identical mean-variance assumption present in the Poisson model by estimating the negative binomial model and testing the null hypothesis that $k = 0$.

Specifications Explored in Protest Count Models

In this analysis, we model as our dependent variable the total number of protests and the number of substantive protests, by both type of contracting center and fiscal year. The protest counts are calculated by counting unique root B-numbers rather than B-numbers.

In our first specification, shown in Table 3.1, we include as our main covariates (1) the natural log of the number of contract awards, (2) the share of awards that are on competitive contracts, (3) the natural log of the average annual expenditure (measured in real terms) per contract, (4) contract center type-specific dummy variables, and (5) a fiscal year time trend.

Table 3.1
Regression Results Under the First Covariate Specification

Specification	OLS		Poisson		Negative Binomial	
	Coef.	Std. Err.	Coef.	Std. Err.	Coef.	Std. Err.
Dependent Variable: Total Protests						
Intercept	186.733**	20.051	176.556**	11.386	178.347**	13.287
Log contract awards	0.763**	0.166	0.665**	0.097	0.685**	0.112
ALC	−0.289	0.336	−0.120	0.227	−0.155	0.252
HQ	−0.029	0.325	0.028	0.210	0.021	0.234
OB	0.342	0.439	0.421	0.286	0.409	0.318
PB	0.214	0.493	0.460	0.313	0.423	0.351
TCC	−0.884	0.477	−0.554	0.308	−0.592	0.344
Fiscal year	−0.093**	0.010	−0.089**	0.006	−0.089**	0.007
Share of awards competitive	−1.170	0.964	−1.424	0.585	−1.439	0.666
Log avg. expenditure/contract/year	−0.122	0.187	−0.033	0.116	−0.038	0.131
Dispersion parameter					0.011	0.009
R-squared	0.819					
Log-likelihood			−268.8		−267.7	
Dependent Variable: Substantive Protests						
Intercept	134.955**	20.827	129.018**	16.893	126.675**	14.659
Log contract awards	0.752**	0.171	0.519**	0.143	0.493**	0.127
ALC	−0.286	0.346	−0.052	0.322	−0.046	0.290
HQ	−0.119	0.334	−0.103	0.303	−0.128	0.272
OB	0.415	0.452	0.351	0.412	0.317	0.368
PB	0.193	0.508	0.403	0.452	0.384	0.401
TCC	−1.138*	0.493	−1.076*	0.455	−1.090**	0.409
Fiscal year	−0.069**	0.011	−0.065**	0.009	−0.064**	0.007
Share of awards competitive	0.142	0.991	−0.736	0.850	−0.820	0.756
Log avg. expenditure/contract/year	−0.043	0.193	−0.018	0.170	−0.029	0.150
Dispersion parameter					−0.017	0.010
R-squared	0.784					
Log-likelihood			−210.8		−209.9	

NOTES: Statistically different from zero with "*" = 95% confidence; "**" = 99% confidence. The excluded contracting center type is PC. The Poisson and negative binomial models include 90 observations each. The OLS regression models for total protests and substantive protests include 89 and 87 observations, respectively. The OLS regressions do not include all 90 observations because those contracting center types with no protests in a particular year must be dropped from the regression.

The second specification, presented in Table 3.2, uses the share of contract awards that are associated with C- and D-type contracts instead of the natural log of the average annual expenditure (measured in real terms) per contract as an alternative measure of contract award size. In the third specification, shown in Table 3.3, we modify the first specification by using the number of competitive awards instead of the number of total awards, hence dropping from our specification the share of awards that are associated with competitive contracts.

Our expectations before estimating the model were that

- The number of protests would rise proportionally with the number of contract awards or number of competitive contract awards, suggesting that the probability of experiencing a protest on a contract award is independent of the number of awards.
- The number of protests would rise with the share of competitive contract awards, holding the total number of contract awards constant. The logic is that there are more offerors, and presumably more opportunities for perceived errors, in competitive source selections than in noncompetitive arrangements.
- The number of protests would rise with the average expenditure per contract or the share of C- and D-type contracts. That is, protests would be more likely with larger awards to the extent that aggregate value is a proxy for contractual complexity.
- The number of protests would be higher in contracting centers with larger and more complex contracts—for example, higher in headquarters and product centers than in projection and other bases.
- The number of protests would fall over time, reflecting the strong relationship identified above when observing trends over time.

We had hoped to further include acquisition workforce information in this analysis. However, lacking data on contractors who support organic personnel assets, we could not construct meaningful measures of the numbers, types, or skills of persons that the Air Force has access to over this period.

Results for Protest Count Regressions

The results of the empirical analysis are presented in Tables 3.1, 3.2, and 3.3. Each table corresponds to a different covariate specification. Analysis of factors that affect the total number of protests versus the number of substantive protests (i.e., protests that lead to corrective actions or merit reviews) yielded qualitatively similar findings. In all of the negative binomial model runs, the parameter k is found not to be statistically different from zero, suggesting that the Poisson model's mean-variance equality assumption is reasonable. Furthermore, the coefficient estimates do not differ greatly between the Poisson and negative binomial models.

The key findings that we robustly observed across all three specifications and modeling approaches are highlighted below.

- All else equal, the total number of protests has fallen at a remarkable 8 to 9 percent per year, even faster than implied by the numbers presented in Chapter Two. The likelihood of a substantive protest being made has been falling at a slightly slower rate of 6 to 8 percent per year.

Table 3.2
Regression Results Under the Second Covariate Specification

Specification	OLS		Poisson		Negative Binomial	
	Coef.	Std. Err.	Coef.	Std. Err.	Coef.	Std. Err.
Dependent Variable: Total Protests						
Intercept	183.914**	36.676	159.842**	22.062	162.646**	25.104
Log contract awards	0.802**	0.155	0.655**	0.094	0.681**	0.108
ALC	−0.128	0.223	−0.059	0.140	−0.092	0.158
HQ	0.171	0.146	0.102	0.108	0.103	0.117
OB	0.617**	0.156	0.518**	0.097	0.517**	0.108
PB	0.510*	0.221	0.579**	0.133	0.550**	0.151
TCC	−0.611**	0.206	−0.513**	0.144	−0.539**	0.159
Fiscal year	−0.093**	0.018	−0.081**	0.011	−0.082**	0.013
Share of awards competitive	−1.193	0.973	−1.356	0.588	−1.373	0.667
Share of awards C- and D-type contracts	0.051	0.375	0.177	0.199	0.173	0.233
Dispersion parameter					0.011	0.009
R-squared	0.818					
Log-likelihood			−268.4		−267.5	
Dependent Variable: Substantive Protests						
Intercept	162.615**	37.585	133.632**	32.287	127.770**	28.607
Log contract awards	0.769**	0.159	0.528**	0.138	0.502**	0.125
ALC	−0.209	0.228	−0.028	0.200	−0.003	0.182
HQ	−0.099	0.149	−0.082	0.152	−0.084	0.141
OB	0.469**	0.159	0.386**	0.138	0.383**	0.125
PB	0.250	0.226	0.436*	0.193	0.452**	0.173
TCC	−0.953**	0.214	−1.023**	0.228	−1.020**	0.214
Fiscal year	−0.083**	0.019	−0.067**	0.016	−0.064**	0.014
Share of awards competitive	0.007	0.994	−0.754	0.859	−0.820	0.764
Share of awards C- and D-type contracts	−0.332	0.383	−0.049	0.299	−0.011	0.256
Dispersion parameter					−0.016	0.010
R-squared	0.786					
Log-likelihood			−210.8		−209.9	

NOTES: Statistically different from zero with "*" = 95% confidence; "**" = 99% confidence. The excluded contracting center type is PC. The Poisson and negative binomial models include 90 observations each. The OLS regression models for total protests and substantive protests include 89 and 87 observations, respectively. The OLS regressions do not include all 90 observations because those contracting center types with no protests in a particular year must be dropped from the regression.

Table 3.3
Regression Results Under the Third Covariate Specification

Specification	OLS		Poisson		Negative Binomial	
	Coef.	Std. Err.	Coef.	Std. Err.	Coef.	Std. Err.
Dependent Variable: Total Protests						
Intercept	168.396**	18.644	159.936**	10.581	161.007**	13.348
Log comp. contract awards	0.752**	0.170	0.712**	0.097	0.726**	0.122
ALC	−0.415	0.339	−0.215	0.227	−0.286	0.269
HQ	−0.173	0.326	−0.072	0.211	−0.102	0.252
OB	0.350	0.451	0.493	0.292	0.461	0.347
PB	0.049	0.500	0.337	0.316	0.259	0.381
TCC	−1.252**	0.458	−0.849**	0.303	−0.943**	0.361
Fiscal year	−0.084**	0.010	−0.081**	0.006	−0.081**	0.007
Log avg. expenditure/contract/year	−0.183	0.190	−0.049	0.119	−0.073	0.144
Dispersion parameter					0.021	0.011
R-squared	0.808					
Log-likelihood			−276.8		−273.6	
Dependent Variable: Substantive Protests						
Intercept	128.519**	18.871	118.301**	15.616	116.803**	14.200
Log comp. contract awards	0.747**	0.171	0.543**	0.143	0.531**	0.131
ALC	−0.329	0.340	−0.114	0.320	−0.107	0.297
HQ	−0.170	0.327	−0.166	0.303	−0.179	0.281
OB	0.420	0.452	0.397	0.418	0.377	0.387
PB	0.134	0.501	0.328	0.455	0.315	0.419
TCC	−1.266**	0.462	−1.261**	0.445	−1.270**	0.414
Fiscal year	−0.065**	0.010	−0.060**	0.008	−0.059**	0.007
Log avg. expenditure/contract/year	−0.065	0.190	−0.030	0.172	−0.035	0.158
Dispersion parameter					−0.013	0.011
R-squared	0.782					
Log-likelihood			−212.3		−211.8	

NOTES: Statistically different from zero with "*" = 95% confidence; "**" = 99% confidence. The excluded contracting center type is PC. The Poisson and negative binomial models include 90 observations each. The OLS regression models for total protests and substantive protests include 89 and 87 observations, respectively. The OLS regressions do not include all 90 observations because those contracting center types with no protests in a particular year must be dropped from the regression.

- The total number of protests, as well as the number of substantive protests, tends to increase as the total number of contract awards rises, but the increase is less than proportional. Alternatively said, the analysis suggests that if the number of contract awards doubles, we would expect to see less than a doubling in protest activity. Because of limitations in the data, we cannot determine what precisely causes the value to be less than one, the value consistent with total proportionality, but the results are robust across functional forms and specifications. Furthermore, this result holds in specification 3 where we use the number of competitive contract awards instead the total number of contract awards as an explanatory variable.

   Note that one apparent cause for this finding is clearly ruled out by our models. When total spending remains constant and the number of contracts rises, dollar value per contract necessarily falls. To the extent that lower value contracts are correlated with lower complexity in terms of the sourcing decision (a common notion in the acquisition community), one would then expect a less-than-proportional increase in protests, as observed here. However, since all three specifications control for the average dollar value per contract, this logic cannot be what underpins the less-than-proportional increase (see the point immediately below). Alternatives such as learning by doing, efficiencies generated by scale, and elevated contractual actions leading to stronger oversight are just some of many potential causes.

- Efforts to identify the effects of larger or more complex acquisitions (through inclusion of expenditure per contract and the share of C- and D-type contracts) found no evidence to suggest that these factors affected the number of protests experienced.

- Contracting centers in headquarters, product centers, and air logistics centers, as defined in Chapter Two, have similar experiences with protests and account for the majority of Air Force procurement. Contracting centers in technical activities, with about 5 percent of the total spending, have substantially better experiences, with between 40 and 71 percent fewer total protests than would be expected in product centers, all else equal.[3] They are likely to have qualitatively different types of contracts and may also sustain a different kind of professional workforce. There are probably useful lessons to be learned from contracting centers in technical activities for the rest of the Air Force; identifying these fell outside the scope of this study. Projection and other bases, with 24 percent of spending, have substantially worse experiences with protest rates than product centers. Although the operational contracts that predominate at these locations are likely simpler, the acquisition workforce is also likely to be less experienced.

---

[3] To derive this range, we first calculated the implied percentage difference in protest rates associated with the technical centers relative to product centers (the excluded contracting center category in the regressions) from each regression. The formula for this calculation is $\exp(b)/\exp(0) - 1$, where b is the coefficient estimate for technical centers. Next, we compare the range of estimates across regressions to identify the highest and lowest estimated effect.

Predicting Protest Outcomes

Another way to test whether various factors have affected Air Force experiences with bid protests is to ask what factors affect the likelihood of a specific outcome when a protest occurs. To do this, we employed multivariate logistic regression.[4] Further, because sustained protests and voluntary corrective actions appear to impose the largest costs on the Air Force when it experiences a protest, we focused our analysis on models that predict the likelihood that a protest is (1) sustained, (2) causes the Air Force to take a corrective action, or (3) is either sustained or causes the Air Force to take a corrective action. We estimate the likelihood of these outcomes for two distinct populations of protests. In particular, we estimate the model for all protests as well as for only those protests that qualify as substantive protests.[5]

The dependent variable of interest here is a dichotomous variable whose value consists of two distinct classes (e.g., whether a protest does or does not result in a particular outcome such as a preemptive corrective action on one or more associated B-numbers). Dummy explanatory variables in the model include the (1) fiscal year of the protest,[6] (2) commodity type of the item or service being procured,[7] (3) protest basis, and (4) contract center type. As is standard, one category for each categorical variable must be excluded from the model. In addition to the categorical variables, the number of B-numbers associated with each root B-number is included as a continuous variable, proxying for the complexity of the protest.[8]

The logistic model implies that explanatory variables relate to the probability of a particular outcome according the following equation:

$$p_i = \frac{e^{\beta x_i}}{1 + e^{\beta x_i}}$$

where p_i is the probability that the ith protest is associated with the outcome under study, β is a vector of coefficients to be estimated, and x_i is a vector of explanatory variables for the ith protest. The model is fitted (i.e., the β vector is estimated) using maximum likelihood.

---

[4] Logistic regression is useful for modeling when the outcome variable of interest is discrete, as is the case here. For an overview of the logistic regression model, see David Hosmer and Stanley Lemeshow, *Applied Logistic Regression*, 2nd ed., John Wiley & Sons, Inc., 2000.

[5] The results are based on 2,482 total protests and 1,059 substantial protests that occurred during FY 1991 to FY 2008 and are reported in PACTS.

[6] We create fiscal year dummy variables for two-year periods instead of one-year periods. We chose to do this to avoid model convergence issues stemming from the fact that no sustained protests occurred during one fiscal year (2004).

[7] Each protested acquisition is categorized into one of five commodity categories based on information in PACTS. The shares of protests occurring in each commodity type category are: construction (10 percent), research and development (R&D) (1 percent), service (58 percent), supply (29 percent), and weapons (2 percent).

[8] For our analysis, we would have liked to have the ability to associate each protest with a measure of the dollar value of the award being protested. Although PACTS provides the solicitation number for the acquisition being protested, we were unable to link the solicitation numbers with contract numbers to identify payments on protested solicitations in the DD350 or FPDS-NG databases. Furthermore, even if we could have linked solicitation and contract numbers, the DD350 database records the values of past transactions, which by themselves do not allow us to calculate the total values relevant to our analysis of ongoing contracts for which we can reasonably anticipate additional transactions. Within the FPDS-NG database, the total value of a contract is estimated only for IDIQ services contracts.

A variety of statistical tests can be used to test the importance of the explanatory variables in predicting the outcome of interest. In particular, we report the chi-squared statistics associated with the likelihood ratio, score, and Wald tests that all βs (except the intercept) equal 0. A p-value of 0.01 for these test statistics suggests that one can reject the null hypothesis that all βs (except the intercept) equal 0 with 99 percent confidence. In addition to these statistics, we report the likelihood ratio tests of the null hypothesis that all coefficients associated with each of the various categorical variables (i.e., fiscal year, commodity type, protest basis, and contracting center type) are equal to zero. This statistical test indicates whether the categorical variable is overall statistically significant.

A positive (negative) coefficient in the logistic model suggests that the variable associated with the coefficient increases (decreases) the probability of the outcome under study. Because we are working primarily with categorical explanatory variables, a positive coefficient means that the associated category increases the probability of the outcome relative to the excluded category. In our output, we indicate which category is excluded from each categorical variable used in the model.

Beyond this, the coefficients from the logistic model are somewhat difficult to interpret. One can exponentiate each coefficient to obtain the variable's odds ratio. The odds ratio represents the ratio of the expected number of times an event will occur over the expected number of times it will not occur. For example, an odds ratio of 4 means that we expect 4 times as many occurrences as non-occurrences. Conversely, an odds ratio of 1/4 means that we expect only one-fourth as many occurrences as non-occurrences. For the categorical variables, the odds ratio should be interpreted relative to the odds ratio of the excluded category.

Logistic Regression Results

Using the logistic regression method described above, we obtain rather modest results. Table 3.4 reports the estimates when the model is run on all protests, and Table 3.5 limits the population used in the estimation to only substantive protests. Summary statistics for each model suggest that our specifications fit the data better than models that include only an intercept term.[9] Our interpretation of the estimates is described below.

- Given either a protest or a substantive protest, the likelihood of a sustained protest did not change over time. That is, fiscal year effects are not correlated with changes in the likelihood of sustained protest outcomes (Models 1 and 4). But fiscal year effects are statistically related to the likelihood of a corrective action (Models 2 and 5).[10] In particular, the fiscal year dummy variable estimates from the corrective action models suggest that during the 1990s, the likelihood of a protest resulting in a corrective action was increasing. This finding is consistent with the upward trend in corrective actions observed during the 1990s. The fact that the fiscal year dummy variables are not found to be a factor in determining sustained protest patterns may result because sustained protests have been too sporadic (and unimportant relative to corrective actions) to identify comparable trends for them.

---

[9] Specifically, in every case, the likelihood ratio, score, and Wald statistics, which are common statistics used to assess model fit, reject the reduced-form model that consists of only an intercept term in favor of our specification.

[10] See the statistics reported in "Likelihood ratio" section of Tables 3.4 and 3.5.

Table 3.4
Logistic Regression Analysis of Protest Outcomes for All Protests

Dependent Variable	Model 1: Sustained Protest		Model 2: Corrective Action Protest		Model 3: Sustained or Corrective Action Protest	
	Coef. Est.	Std. Err.	Coef. Est.	Std. Err.	Coef. Est.	Std. Err.
Intercept	(4.71)**	0.86	(1.97)**	0.484	(1.78)**	0.43
Fiscal year (excluded year = 2007–2008)						
1991–1992	0.44	0.60	(2.38)**	0.34	(2.03)**	0.29
1993–1994	0.15	0.56	(1.03)**	0.21	(0.96)**	0.21
1995–1996	(0.31)	0.62	(0.83)**	0.22	(0.84)**	0.22
1997–1998	(0.02)	0.63	(0.08)	0.22	(0.10)	0.22
1999–2000	0.26	0.62	(0.16)	0.23	(0.21)	0.23
2001–2002	(0.27)	0.71	0.16	0.23	0.01	0.23
2003–2004	(0.69)	0.79	0.07	0.23	(0.08)	0.23
2005–2006	1.11*	0.56	0.22	0.23	0.34	0.23
Commodity (excluded commodity = weapons)						
Construction	(0.53)	0.89	(0.38)	0.40	(0.55)	0.39
R&D	0.19	1.06	(0.87)	0.63	(1.00)	0.61
Service	0.20	0.67	(0.31)	0.36	(0.44)	0.35
Supply	(0.57)	0.71	(0.03)	0.37	(0.24)	0.36
Protest basis (excluded protest basis = other)						
Faulty evaluation	1.17	0.42	0.40*	0.18	0.37*	0.17
Faulty RFP	0.30	0.44	0.78**	0.19	0.72**	0.18
Faulty sourcing decision	0.81*	0.41	0.18	0.20	0.16	0.20
Faulty treatment of offer	(0.08)	0.50	1.00**	0.20	0.82**	0.19
Contract center type (excluded contract center type = PC)						
ALC	(0.07)	0.39	0.29	0.22	0.28	0.21
HQ	(0.69)	0.48	0.30	0.24	0.16	0.23
OB	(0.28)	0.40	0.35	0.22	0.28	0.21
PB	(1.58)**	0.47	0.53**	0.20	0.35	0.19
TCC	(2.04)	1.05	0.21	0.30	(0.00)	0.29
Number of B-numbers	0.37**	0.08	0.32**	0.06	0.50**	0.06
Diagnostic statistics	Chi-square	p-value	Chi-square	p-value	Chi-square	p-value
Test that all B = 0						
Likelihood ratio	110.60	<.0001	267.44	<.0001	303.57	<.0001
Score	183.10	<.0001	255.94	<.0001	293.45	<.0001
Wald	90.71	<.0001	203.17	<.0001	228.92	<.0001

Table 3.4—Continued

Dependent Variable	Model 1: Sustained Protest		Model 2: Corrective Action Protest		Model 3: Sustained or Corrective Action Protest	
	Coef. Est.	Std. Err.	Coef. Est.	Std. Err.	Coef. Est.	Std. Err.
Intercept	(4.71)**	0.86	(1.97)**	0.484	(1.78)**	0.43
Likelihood ratio test of qualitative factors						
Fiscal year	13.70	0.09	119.11	<.0001	113.45	<.0001
Commodity	5.93	0.20	7.74	0.10	6.44	0.17
Protest basis	9.62	0.05	38.17	<.0001	28.85	<.0001
Contract center type	17.33	0.00	8.36	0.14	5.02	0.41
Summary statistics						
Observations		2,482		2,482		2,482
Number of observations with dependent variable = 1		70		533		590

NOTE: Statistically different from zero with "*" = 95% confidence; "**" = 99% confidence.

- The basis for protest is correlated with the likelihood of a corrective action, but identifying the exact relationship is challenging. Given a protest, claims of a faulty RFP or faulty treatment of an offeror appear somewhat more likely than other claims to be associated with a corrective action. Given a substantive protest, only the claim of a faulty RFP stands out as exceptional in its effect on the likelihood of a corrective action.

  The effect of the basis of protest on the likelihood of a sustained protest is even more opaque. Given a protest, a sustained protest was somewhat more likely when the basis of protest involved a faulty sourcing decision. Yet given a substantive protest, any such effect disappears. This suggests that a faulty sourcing decision may have increased the likelihood of a substantive protest; stated differently, protesters have been least likely to file a protest in error when claiming a faulty sourcing decision.
- Given either a protest or a substantive protest, the type of commodity bought had no discernible relationship with the likelihood of a sustained protest or a corrective action in all but one specification.
- The likelihood test indicates that contracting center type has a statistically significant relationship with whether GAO sustains a protest (both Models 1 and 4), but it is generally not possible to statistically differentiate between the effects of any two types of contracting centers. Alternatively, the specific contracting center type is statistically related to a corrective action outcome when the population is limited to only substantive protests (Model 5).
- Protests associated with multiple B-numbers are more likely to be associated with a sustained ruling or a corrective action with the exception of Model 5 (corrective actions for substantive protests). The model coefficient of 0.37 estimated on the B-number variable in both Models 1 and 4 suggests that the odds of a protest receiving a sustained ruling from GAO increases by 1.45 (= exp(0.37)) for every additional B-number associated with a protest. In other words, the odds of receiving at least one sustained ruling on at least one protest B-number are 45 percent higher for a protest with two B-numbers as opposed to one.

Table 3.5
Logistic Regression Analysis of Protest Outcomes for Substantive Protests

Dependent Variable	Model 4: Sustained Protest		Model 5: Corrective Action Protest		Model 6: Sustained or Corrective Action Protest	
	Coef. Est.	Std. Err.	Coef. Est.	Std. Err.	Coef. Est.	Std. Err.
Intercept	(3.63)**	0.91	0.37	0.55	0.76	0.58
Fiscal year (excluded year = 2007–2008)						
1991–1992	0.60	0.63	(3.14)**	0.42	(2.89)**	0.41
1993–1994	0.21	0.58	(1.94)**	0.33	(2.00)**	0.35
1995–1996	(0.39)	0.64	(1.71)**	0.33	(1.89)**	0.36
1997–1998	(0.22)	0.65	(0.68)*	0.34	(0.82)*	0.36
1999–2000	0.09	0.65	(1.15)**	0.35	(1.36)**	0.37
2001–2002	(0.60)	0.73	(0.57)	0.36	(0.94)*	0.38
2003–2004	(1.07)	0.80	(0.81)*	0.35	(1.22)**	0.37
2005–2006	0.85	0.59	(0.63)	0.35	(0.51)	0.39
Commodity (excluded commodity = weapons)						
Construction	(0.25)	0.90	0.04	0.48	(0.11)	0.49
R&D	0.23	1.11	(0.99)	0.68	(1.18)	0.68
Service	0.32	0.68	0.04	0.42	(0.06)	0.44
Supply	(0.43)	0.72	0.47	0.43	0.25	0.45
Protest basis (excluded protest basis = other)						
Faulty evaluation	0.53	0.44	(0.23)	0.23	(0.22)	0.24
Faulty RFP	0.16	0.46	1.06**	0.25	1.08**	0.25
Faulty sourcing decision	0.62	0.43	0.14	0.26	0.18	0.26
Faulty treatment of offer	(0.48)	0.50	0.44	0.24	0.24	0.24
Contract center type (excluded contract center type = PC)						
ALC	0.08	0.40	0.32	0.25	0.33	0.25
HQ	(0.59)	0.50	0.46	0.29	0.28	0.28
OB	(0.05)	0.42	0.50	0.26	0.45	0.26
PB	(1.37)**	0.49	0.93**	0.24	0.68**	0.24
TCC	(1.57)	1.07	0.80*	0.37	0.52	0.36
Number of B-numbers	0.37**	0.08	0.05	0.06	0.20**	0.06
Diagnostic statistics	**Chi-square**	**p-value**	**Chi-square**	**p-value**	**Chi-square**	**p-value**
Test that all B = 0						
Likelihood ratio	81.36	<.0001	191.47	<.0001	172.37	<.0001
Score	95.27	<.0001	175.10	<.0001	159.93	<.0001
Wald	66.65	<.0001	146.24	<.0001	137.80	<.0001

Table 3.5—Continued

Dependent Variable	Model 4: Sustained Protest		Model 5: Corrective Action Protest		Model 6: Sustained or Corrective Action Protest	
	Coef. Est.	Std. Err.	Coef. Est.	Std. Err.	Coef. Est.	Std. Err.
Intercept	(3.63)**	0.91	0.37	0.55	0.76	0.58
Likelihood ratio test of qualitative factors						
Fiscal year	14.33	0.07	101.82	<.0001	94.15	<.0001
Commodity	4.99	0.29	11.52	0.02	8.74	0.07
Protest basis	4.48	0.34	45.00	<.0001	42.02	<.0001
Contract center type	14.26	0.01	19.31	0.00	9.56	0.09
Summary statistics						
Observations		1,059		1,059		1,059
Number of observations with dependent variable = 1		70		533		590

NOTE: Statistically different from zero with "*" = 95% confidence; "**" = 99% confidence.

The lack of correlation between the type of commodity and protest outcome and the opaque results for the basis of protest may to a degree reflect the quality of data in PACTS. Our review of the quality of PACTS data on both of these factors detected inconsistencies and irregularities that forced us to scrub the data thoroughly before use. For example, ways of defining the basis of protest appear to have changed over time. Moreover, it is unclear how the person providing data to PACTS decided which label, from a list of basis of protest, to use for any one protest. Similar concerns apply to the data on the type of commodity. Therefore, we report these findings with some caution and suggest that the effects of basis of protest and type of commodity would benefit from additional attention in the event improved data become available.

CHAPTER FOUR

Summary and Conclusions

Although public and congressional criticism of the Air Force has emphasized recent protests sustained by GAO, they are only part of a much larger picture. When a protest occurs, the Air Force can offer preemptive corrective action of its own accord. That is, the Air Force can offer to reevaluate proposals submitted, reopen evaluations, and give offerors an opportunity to adjust their proposals; change the offerors included in the source selection; rewrite the request for proposals and start the source selection from scratch; or even choose to cancel the solicitation process as a whole. In the event a protestor accepts such action, there is no further need for GAO to review the protest. Between FY 2000 and FY 2008, the Air Force experienced 836 protests. It offered corrective actions in 273 instances, or 33 percent of the time. Meanwhile, it ultimately suffered sustained protests in only 29 instances, or 3 percent of the time.

To put protests in perspective, it is useful to express them in terms of the total number of contract awards the Air Force makes. The number of protests as a percentage of total contract awards has fallen fairly steadily from about 1.7 percent in FY 1995 to 0.5 percent in FY 2008. Sustained protests are so unusual that they hardly register relative to the total number or value of contract awards. On average, between FY 2000 and FY 2008, GAO sustained one protest for every $20 billion the Air Force spent in acquisitions. So few sustained protests have occurred that it is impossible to discern any trend in them. On the other hand, the Air Force has offered corrective actions in noticeable numbers. Through the 1990s, it offered corrective actions in about 0.3 percent of contract awards. From FY 2001 to FY 2008, the percentage has fallen fairly steadily, ending well under 0.2 percent of contract awards. All of these trends point to steady improvement over time, resulting in a need to adjust fewer than 0.2 percent of the source selections associated with contract awards by FY 2008.

In this technical report, we provide tabulations of data from administrative databases that support these findings. We also perform statistical analyses to understand the factors correlated with protest activity and outcomes, controlling for other factors that may be attributing to protests. The statistical analyses suggest the following additional findings:

- After controlling for other factors, the likelihood of receiving a protest has been declining at a rate of 8 to 9 percent per year from FY 1994 to FY 2008. When attention is restricted to the likelihood of receiving a protest that undergoes a merit review by GAO or receives a corrective action, the rate of decline falls slightly but still exceeds 6 percent per year in most of the empirical specifications we tested. This finding is consistent with general trends observed in the protest record but also controls for other factors that are likely to be important, such as the amount of contracting activity being undertaken by the Air Force.

- Protest activity tends to increase as the total number of contract awards rises, but the increase is less than proportional. The analyses suggest that if the number of contract awards doubles, we would expect to see less than a doubling in protest activity, holding all other factors equal including expenditures. These results are robust across models.

- When assessing protest performance across Air Force types of contracting centers, contracting centers in technical activities—those locations that conduct test and evaluation activities—tend to perform better than other types of Air Force contracting centers, holding constant such other important factors as spending and the number and nature of contracts. The acquisition environment in these centers (e.g., personnel, policies) can likely provide useful lessons for the rest of the Air Force.

- In terms of predicting protest outcomes based on the characteristics of the underlying protest (i.e., whether a protest is sustained by GAO or results in a corrective action by the Air Force), our statistical findings tend to be less robust, but some findings are worth briefly noting. Namely, assessment of protest outcome determinants supports the casual observation that over the 1990s there was a steady increase in the probability that a protest received a corrective action, while controlling for other trends that were also occurring at that time. This said, we did not find a significant correlation between protest outcomes and the basis for a protest (e.g., problems with the request for proposal, faulty proposal evaluations) or the commodity or service being acquired by the Air Force. Furthermore, there is some evidence that when taken as a whole, different types of contracting centers are associated with different probabilities of experiencing a sustained ruling by GAO or leading to Air Force corrective actions; however, it is difficult to statistically differentiate between any two types of centers.

Data Used in the Analyses

The analyses conducted in Chapters Two and Three rely on data collected on Air Force acquisition protests contained in the PACTS database as well as procurement data that were tabulated from the DD350 and FPDS-NG databases. In this appendix, we describe these databases and how they were used in the analysis.

Protest Data (PACTS)

The PACTS database is maintained and updated by the Secretary of the Air Force for Acquisition (SAF/AQC) as a repository of information on bid protests of Air Force contract awards that have been submitted to GAO. The database includes information on acquisition protests submitted to GAO against the Air Force over the FY 1991 to FY 2008 period. A copy of the PACTS database was provided to RAND on February 28, 2009. Protests reported in PACTS are limited to those involving Air Force acquisitions. As a result, our analyses of protest activity using PACTS do not integrate information on protests against other services.

To improve our understanding of the PACTS data, we spoke to the principals responsible for managing and updating the database. They provided a useful corporate memory about significant past experience with the data system and their assessment of the reliability of certain information contained in the current version of the database.

Over the FY 1991 to FY 2008 period, the way data have been entered and updated in PACTS has evolved. In particular, additional fields have been added and the categorization of protest attributes and outcomes has changed in some instances. In some of our analyses, our focus has been limited to the post-FY 1999 period, because data on fields of interest were either not recorded or were recorded differently during earlier years. We applied considerable effort to understand these changes and "clean" the PACTS data so that valid inferences on protest trends could be made.

PACTS Database Structure and Fields

When an offeror files a protest, GAO assigns the protest one or more B-numbers (i.e., B-123456.2, B-123456.3). GAO may assign multiple B-numbers to a source selection protest if multiple offerors file protests or the protested actions encompass distinct issues as defined by GAO. B-numbers associated with the same acquisition will generally have the same root

B-number (i.e., B-123456) under the GAO system, although some exceptions to this do exist. A record in PACTS is created for each protest B-number that the Air Force tracks.

Past analyses of protest trends have typically relied on counts of B-numbers.[1] Because a protest can be associated with one or more B-number, these analyses typically overstate the true number of protests. Counts of root B-numbers are more indicative of the number of acquisitions that have been protested. In the analyses presented in this technical report, tabulations of the number of protests correspond to counts on unique root B-numbers unless specifically noted otherwise.

During the period FY 1991 to FY 2008, the PACTS data provided to RAND had 4,104 observations of unique B-numbers that are associated with 2,864 unique root B-numbers. For the more recent period of FY 2000 to FY 2008, the PACTS data had 1,260 B-numbers that correspond with 836 unique root B-numbers.

Table A.1 describes the key fields that we used in PACTS. The quality and consistency of data presented in these fields differ significantly. For some fields, we know that the way data are reported has changed over time. In these cases, we used data from multiple fields to reclassify data in a more consistent way. In particular, we took the following data cleaning and categorizing steps:

1. PACTS does not include a record of root B-numbers. We derived root B-numbers from B-numbers by rounding down each B-number to a whole number (i.e., we associated

Table A.1
Key Fields from PACTS Used in RAND Protest Analyses

PACTS Field Name	Field Description
Air Force base	Air Force Base where the procuring agency is located
B number	Protest B-number
Commodity	Classifies the commodity or service purchased using commodity type codes
Corr Actn Taken	Indicator variable equal to 1 if a corrective action was taken, 0 otherwise
Corr Actn Type	Description of the type of corrective action taken
Decision Type	Description of the reason for a GAO decision. Used to analyze why protests were dismissed by GAO
Fiscal Year	Fiscal year in which the protest occurred
GAO Decision	Indicates whether the protest was dismissed, denied, etc.
Item or Service	Detailed description of the item or service procured
MAJCOM	MAJCOM of the procuring agency
Office Symbol	Office symbol of the procuring agency
Protest Basis	Description of the basis for protest using a discrete list of protest basis codes
Protest Text	Detailed description of protest basis

---

[1] See, for example, Congressional Research Service, 2009a. This is discussed in Congressional Research Service, 2009b.

the B-number 123456.7 with the root B-number 123456). We performed this step for every observation in the PACTS database.

2. To conduct the analyses at the root B-level, we "rolled up" the PACTS data to create a new analytical dataset in which each observation represents a unique root B-number (rather than a unique B-number).

3. We assigned each PACTS observation to one of six types of contracting centers based on the MAJCOM, Air Force base, and office symbol of the contracting office. Chapter Two describes the exact criteria we used to associate protests and acquisitions with contracting center types.

4. We assigned each observation to one of five commodity or service types. To do this, we first relied on the "Commodity" field in PACTS. If the observation had a missing value, we then relied upon the "Item or Service" field.

5. PACTS provides information on the reason for a protest by specifying one of 30 protest basis categories in the "Protest Basis" field. Reporting of protest basis seems to have changed over time, with some categories being used much more or less frequently in earlier or later periods. To facilitate our analyses and discern trends, we grouped the 30 protest basis categories into six classifications. For about 10 percent of the sample where no protest basis was listed in PACTS, we relied on the "Protest Text" field to classify the protest into one of our six protest basis groupings.

6. For protests that led the Air Force to take a corrective action, PACTS reports the type of corrective action taken. The classification scheme applied appears to have evolved over time. To create consistency in our analysis of corrective actives, we classified each type of corrective action reported in PACTS into one of six broad categories based on the information available in the "Corr Actn Type" field.

7. The reason for a GAO decision reported in PACTS also seemed to suffer from an inconsistent categorization over time. This information is contained in the "GAO Decision" field within PACTS. To facilitate our analyses, we recategorized the reasons for a GAO decision into a manageable list of broad GAO decision reasons based on information contained in the "GAO Decision" field.

In addition to these steps, we made some effort to validate the data contained in PACTS. We drew a random sample of 100 root B-numbers from PACTS, which corresponded to 139 protest B-numbers. Next, we searched GAO's website for protest decisions covering the B-numbers in the sample. We found 46 decisions covering 50 B-numbers (or approximately 36 percent of the sample) on GAO's website. Personnel in the Air Force and GAO told us that GAO's website is incomplete, so the low match rate does not necessarily indicate an issue with PACTS.

We compared information contained in PACTS and the GAO decisions. This review indicates the following:

- In every case that could be verified, the decision outcome (i.e., dismissed, sustained), Air Force base, and item or service listed in PACTS matched that reported in the GAO decision.
- The decision date reported in the GAO decision differed from that reported in PACTS in four out of 50 cases. In these four cases, the decision date differed from as little as one

day to as many as 61 days. These discrepancies appear minor, given that our analyses are based only on the fiscal year of the protest.

- The basis for protests reported in PACTS largely coincides with the discussion of the protest reasons described in the GAO decision.
- In four decisions of the 46 sampled, four B-numbers not reported in PACTS were included in the GAO decisions. These decisions were written in FYs 1991, 1993, 1994, and 1997. In one case, the B-number not referenced in PACTS was associated with a "claim for reimbursement . . . of legal fees incurred in connection with winning protest issue." In another case, the B-number not found in PACTS was tied to a "request for reconsideration" of a previous protest. In the other two cases, the exact protest reason is unclear because the decision covered multiple B-numbers. Personnel at SAF/AQC told us that PACTS does not systematically record subsequent protests based on a request for legal fees or request for reconsideration.

Although this review does not constitute a formal validation of the PACTS data, it does suggest that PACTS is a useful source of information on protests. In particular, in many cases, we found that the PACTS data contained more information than is available in the GAO decision. In a few instances, we found discrepancies between the two sources of data, although we do not believe that these discrepancies are large enough to invalidate the PACTS data or the inferences that we have drawn from them.

Procurement Data (DD350 and FPDS-NG)

DD350 and FPDS-NG databases are the official data sources for overall detailed DoD spending.[2] Within the DoD, until FY 2007, procurement transaction information was traditionally reported on DoD Form 350, the Individual Contract Action Record, and known as DD350 data. At the end of each fiscal year, OSD collected and scrubbed these data and posted a final database to its website. Beginning in FY 2007, these transaction data are recorded through automated interfaces to the FPDS-NG. Both sources of data include information on "contract actions." This information includes the name of the contract and contracting office associated with each action, the change in federal financial obligations associated with the action, and other related information.

Past RAND studies have analyzed DD350 and FPDS-NG data on spend analyses for the Air Force and the Office of Small Business Programs.[3] Using our data from FY 1994 to FY 2006, we have been able to replicate the total dollars and actions reported as SIAD's official

---

[2]   The Federal Funding Accountability and Transparency Act of 2006 (Transparency Act) requires a single searchable website, accessible by the public for free. The website created, called USAspending.gov/, uses FPDS-NG data. FPDS-NG data can be queried directly or downloaded from the FPDS-NG home website. DD350 data can be downloaded from the DoD Procurement web page of the Statistical Information Analysis Division (SIAD), Office of the Secretary of Defense.

[3]   Lloyd Dixon, Chad Shirley, Laura H. Baldwin, John A. Ausink, and Nancy Campbell, *An Assessment of Air Force Data on Contract Expenditures*, Santa Monica, Calif.: RAND Corporation, MG-274-AF, 2005; Nancy Y. Moore, Clifford A. Grammich, Julie S. DaVanzo, Bruce J. Held, John Coombs, and Judith D. Mele, *Enhancing Small-Business Opportunities in the DoD*, Santa Monica, Calif.: RAND Corporation, TR-601-1-OSD, 2008.

totals for each fiscal year. No official numbers exist for FY 2007 to FY 2008 FPDS-NG data, because contract actions continue to be added to these databases over time.[4]

In an effort to put the protest data in perspective, we used the DD350 and FPDS-NG data in this study to characterize procurement activity over time and across types of Air Force contracting centers. We limited our use of these data to FY 1994 to FY 2008, because the contracting office code names for Air Force contracting offices in the FY 1991 to FY 1993 data had Army and Navy contracting office names mixed in. Extensive data scrubbing would have been required to remove these non–Air Force contract actions.

DoD policy required that any contract action equal to or greater than $25,000 in absolute value and measured in current-year dollars be recorded in the DD350 system. These data record both obligations and deobligations.[5] Beginning in FY 2007, the threshold of contract actions dropped to $2,500. Thus, FPDS-NG records all actions that are equal to or greater than $2,500 in absolute value. In an effort to treat the data consistently over time, for the entire study period of FY 1994 to FY 2008, we include only contract actions at or above $25,000 in absolute value.

However, we were able to associate contract expenditures with fiscal years and contracting centers.[6] We grouped contracting offices for each contract action by MAJCOM, Air Force base, and office symbol and then assigned each to one of the six types of contracting centers outlined in Chapter Two. We expected that some contracting organization types would award contracts that were more similar in value and type as a group than other organizations. We hypothesized that product centers typically award larger contracts than air logistics centers. In addition, we reasoned that some organizations awarded contracts that were qualitatively different from other Air Force organizations. Thus, testing and research and development labs might award different kinds of contracts than operating bases and could experience different rates of bid protests.

With the above rationale in mind, we were able to test whether certain kinds of Air Force organizations experienced more sustained bid protests or corrective actions than others. We made tabulations by contracting center type and fiscal year of the total number and the associated dollars for (1) all contract actions; (2) all new contract awards; (3) all new competitive contract awards; (4) all new large contract awards—i.e., indefinite quantity/indefinite delivery and definite quantity/indefinite delivery contracts;[7] and (5) all new competitive large contract awards.

---

[4] The FY 2007 Air Force FPDS-NG data were downloaded from the FPDS-NG website on January 10, 2008. The FY 2008 Air Force FPDS-NG data were downloaded on April 18, 2009.

[5] An *obligation* here represents a contracted financial obligation. A *deobligation* represents a downward adjustment of the obligations recorded in a contract document and can be caused by factors such as (1) termination of a part of the project, (2) reduction in material prices, (3) cost underrun, or (4) correction of recorded amounts.

[6] Ideally, to analyze whether the expected size of a contract award affects the likelihood of a bid protest, we would have liked to identify the total dollar value of each contract award. The FPDS-NG data records ultimate contract value, which is the total estimated value of the contract and all of its options over the life of the contract. However, since data for the FY 1994 to FY 2006 period came from DD350, which does not record ultimate contract value, we did not have a comparable measure for this time period. We use average contract expenditures per contract by fiscal year and contracting center as a proxy for contract value on awarded contracts in the protest count analysis presented in Chapter Three.

[7] Indefinite quantity/indefinite delivery contracts have a "D" in the 9th position of their contract numbers. Definite quantity/indefinite delivery contracts have a "C" in the 9th position of their contract numbers. Large dollar contracts of the sort of most interest to the study's sponsor are usually one of these types of contracts.

Table A.2 describes the data elements analyzed in DD350 and FPDS-NG data. We defined the contract number to be the first 13 positions of the basic contract number. We used the "extent competed" variable to determine whether the contract was competitive. Obligations and deobligations provided dollar values. "Number of actions" indicated the number of actions per data record. More than one action can be recorded in a single observation. The contract office code denotes the office that initiated the contract action. This action could occur in the context of a contract written by the purchasing organization or others, even non-Air Force organizations, such as the Defense Logistics Agency or the General Services Agency. We made contracting center classifications by attaching contract office code names and addresses from files available from the DD350 and FPDS-NG websites.

Some of the statistical analyses used data on number of contracts and contract actions by contracting center type for each fiscal year. In this case, we analyzed all contracts and associated actions for each fiscal year they were recorded. We also summed their associated dollars and number of actions. Other analyses considered contract awards. In this case, we assigned contract numbers to the fiscal year and contracting center when the first action was recorded and then analyzed awards accordingly. Thus, the award of a five-year contract would be counted in the first year it became active with the first contracting center recorded but not in the ensuing four years or for any other contracting centers. This is because a contract award occurs only once.

Table A.2
Key Fields from DD350 and FPDS-NG Data Used in RAND Bid Protest Analyses

	Variable Names	
Variable Description	**DD350**	**FPDS-NG**
Fiscal year	FY of the file	FY of the file
Reporting agency	A3A Reporting Agency	agencyID
Contract number	B1A Contract Number	PIID (procurement instrument identifier)
Type of contract	9th position of Contract Number	9th position of PIID
Competitive	C3 Extent Competed	extentCompeted
Expenditure in dollars	B8 Obligated or Deobligated Dollars	obligatedAmount
Number of actions	E5 Number of Actions	numberOfActions
Contract office code	A3B Contracting Office	contractingOfficeID
Contract office name	CONTRACTING_OFFICE_NAME	CONTRACTING_OFFICE_NAME
Base or city	ADDRESS_CITY	ADDRESS_CITY
State	ADDRESS_STATE	ADDRESS_STATE
Contracting center type	1 of 6 different types of Air Force organizations	1 of 6 different types of Air Force organizations
Fiscal year when first awarded	FY associated with the first transaction for each contract number	FY associated with the first transaction for each contract number
Fiscal year when first competed	FY associated with the first transaction for each competitive contract	FY associated with the first transaction for each competitive contract

Bibliography

Bureau of Economic Analysis, GDP Data, undated. As of February 11, 2010:
http://www.bea.gov/national/index.htm

Camm, Frank, Mary E. Chenoweth, John C. Graser, Thomas Light, Mark A. Lorell, Rena Rudavsky, and Peter Lewis, *Government Accountability Office Bid Protests in Air Force Source Selections: Evidence and Options,* Santa Monica, Calif.: RAND Corporation, DB-603-AF, 2012a. As of January 24, 2012:
http://www.rand.org/pubs/documented_briefings/DB603.html

Camm, Frank, Mary E. Chenoweth, John C. Graser, Thomas Light, Mark A. Lorell, and Susan K. Woodward, *Government Accountability Office Bid Protests in Air Force Source Selections: Evidence and Options—Executive Summary,* Santa Monica, Calif.: RAND Corporation, MG-1077-AF, 2012b. As of January 24, 2012:
http://www.rand.org/pubs/monographs/MG1077.html

Congressional Research Service, *GAO Bid Protests: Trends, Analysis, and Options for Congress,* Washington, D.C., February 11, 2009a.

Congressional Research Service, *Report to Congress on Bid Protests Involving Defense Procurements,* B-401197, Washington, D.C., April 9, 2009b.

Dixon, Lloyd, Chad Shirley, Laura H. Baldwin, John A. Ausink, and Nancy Campbell, *An Assessment of Air Force Data on Contract Expenditures,* Santa Monica, Calif.: RAND Corporation, MG-274-AF, 2005. As of November 1, 2010:
http://www.rand.org/pubs/monographs/MG274.html

Greene, William, *Econometric Analysis,* 6th ed., Pearson–Prentice Hall, 2008.

Hosmer, David W., and Stanley Lemeshow, *Applied Logistic Regression,* 2nd ed., John Wiley & Sons, Inc., 2000.

Kayes, Brett N. (Capt, USAF), "Air Force GAO Protest Trend Analysis," briefing, Washington, D.C.: SAF/AQC, updated September 19, 2008.

Moore, Nancy Y., Clifford A. Grammich, Julie S. DaVanzo, Bruce J. Held, John Coombs, and Judith D. Mele, *Enhancing Small-Business Opportunities in the DoD,* Santa Monica, Calif.: RAND Corporation, TR-601-1-OSD, 2008. As of November 1, 2010:
http//:www.rand.org/pubs/technical_reports/TR601-1.html